MIRAGE

MIRAGE

FLORIDA AND THE VANISHING WATER OF THE EASTERN U.S.

Cynthia Barnett

THE UNIVERSITY OF MICHIGAN PRESS ANN ARBOR

First paperback edition 2008

Published in the United States of America by
The University of Michigan Press
Manufactured in the United States of America
♾ Printed on acid-free paper

2011 2010 2009 2008 6 5 4

A CIP catalog record for this book is available from the British Library.

Library of Congress Cataloging-in-Publication Data

Barnett, Cynthia, 1966–
Mirage : Florida and the vanishing water of the Eastern U.S. /
Cynthia Barnett.
p. cm.
Includes bibliographical references and index.
ISBN-13: 978-0-472-11563-1 (cloth : alk. paper)
ISBN-10: 0-472-11563-4 (cloth : alk. paper)
1. Water consumption—Florida. 2. Water consumption—East (U.S.)
3. Florida—Environmental conditions. 4. East (U.S.)—
Environmental conditions. 5. Water conservation—Florida. 6. Water
conservation—East (U.S.) 7. Water-supply—Florida. 8. Water-
supply—East (U.S.) I. Title.

TD224.F6B368 2007
333.91'1309759—dc22 2006036415

ISBN-13: 978-0-472-03303-4 (pbk. : alk. paper)
ISBN-10: 0-472-03303-4 (pbk. : alk. paper)

A Caravan book. For more information, visit www.caravanbooks.org.

TO MY GRANDMOTHERS. *Elsie Catherine Taylor Barnett* in childhood walked through fragrant orange groves to pump drinking water for her family in Pinellas County on Florida's west coast. Now the densest county in the state, Pinellas has no remaining fresh groundwater supply. On the opposite coast of Florida, *Nancy Ann Lindboe Drews* galloped her palomino through the wet jungles of Broward County. They are now urban jungles.

CONTENTS

Prologue 1

1. History and Myth 12
2. Conspicuous Consumption 31
3. Drained and Diverted 44
4. The Wetlands and the Weather 59
5. Red State, Green State 77
6. Destination: Florida 98
7. Water Wars 114
8. Business in a Bottle 128
9. Priceless 145
10. Water Wildcatters 157
11. Technology's Promise 168
12. Redemption and the River of Grass 180

Acknowledgments 193 *Notes* 197
Bibliography 219 *Index* 225

Prologue

THE CRACK CREPT just like ivy. It sprouted from below ground, then inched up the brick of David and Vivian Atteberry's home in an Orlando, Florida, suburb. With a thick, black marker, David Atteberry measured its journey, along with those of the cracks that had appeared in the ceilings and at the edges of almost every window in the house.

When the crack in the brick grew six inches in one day, David Atteberry called his insurance company. The adjuster came to see, and called a geologist. The geologist drilled a hole, and left in a hurry.

"Mr. Atteberry, are you sitting down?" the adjuster asked when she called.

He sat.

She told him that the ground was swallowing his house.

"It's a massive sinkhole," she said. "You should pack your family and get out of there."[1]

Back home in Illinois, the Atteberrys had never heard of sinkholes. In Florida, they are common enough: some too small to notice, some big enough to sink a chunk of highway or a Porsche dealership, as one did in Winter Park in 1981.

Sinkholes are collapses in the limestone rock that underlies Florida.

The peninsula sits atop what geologists call "karst," a pocked terrain formed over millions of years as water dissolved the limestone to create sinkholes, as well as Florida's spectacular blue springs and its mysterious underground rivers and caves.

These shifting "sinks," as they are known, are as natural to Florida as the waves that shape the state's 1,400-mile coastline. But human activity can open them up, too: highway construction, excavation of fill dirt, well drilling, and, particularly, the excessive pumping of groundwater.[2]

In the last half century, Florida has seen extraordinary population growth—from 2.8 million people in 1950 to 17 million today. The current decade will bring about "the largest absolute population increase of any decade in Florida's history," says Stanley K. Smith, director of the University of Florida's Bureau of Economic and Business Research. Florida has a net influx of 1,060 people every single day. The math looks like this: 1,890 move in; 945 move out; births outnumber deaths by 115; total average daily population growth equals 1,060.[3]

Among obvious consequences like traffic gridlock and crowded schools, this relentless growth causes thousands of other, more subtle problems, for one, an increase in the severity and frequency of sinkholes. To supply water to more than 90 percent of its booming population, Florida relies on groundwater pulled up from permeable aquifers underground. Almost everywhere else in the United States, water withdrawals have flattened in recent years despite population growth, thanks to conservation and greater efficiencies in water use. But the Sunshine State sucks up more and more water all the time, primarily to keep its fast-spreading lawns and golf courses green.

Today, Floridians are pumping groundwater out of their aquifers faster than the state's copious rainfall can refill them. Meanwhile each new master-planned community, shopping mall, and highway drains water in a bit of a different direction and lowers groundwater levels a little bit more. These are precisely the sorts of geologic disturbances that cause sinks, essentially funnels in the porous limestone.

In Central Florida, the sinkhole problem has become prevalent enough that the government saw fit to put out a brochure for homeowners. Called "Sinkholes," its cover shows a single-family home half-toppled into a huge crater of sand and water. The booklet pinpoints the most sinkhole-prone part of Florida, a stretch of the central west coast that draws blue-collar retirees who have cashed out of the Midwest to afford a modest home in a planned community. The most sinkhole-vulnerable county, Pasco, is also one of the one hundred fastest-growing counties in the United States.[4]

It would be handy to come across the brochure while house hunting. In Florida, that is about as likely as finding a real estate Web site with a link to the National Hurricane Center. But soon, home buyers will learn about sinkholes: when they are denied insurance. Florida's major carriers have quit writing policies in those parts of the state where sinkhole claims are highest.

Crisis? Families like the Atteberrys would say so. But sinkholes are just one small symptom of a much greater problem facing Florida and other parts of the eastern United States for the first time since humans began living here some 12,000 years ago.

A shortage of life's most important ingredient. Water.

Until recently, people in the eastern United States enjoyed an abundance of freshwater. In fact, they thought there was far too much of it. In 1876, Major John Wesley Powell, the adventuresome one-armed explorer who then headed the U.S. Geological Survey, declared that a longitudinal line along the 100th meridian, down the middle of North and South Dakota, Nebraska, Kansas, Oklahoma, and Texas, divided a moist East from an arid West. To the west of the line, he reported to Congress, a lack of rainfall would require cooperative irrigation and an equitable system of water rights to ensure scarce water would be used for the greatest good. To the east of the line, more than 20 inches of rainfall a year meant that people could settle and grow anything they wanted without irrigation.

Powell, the first American to explore the wild Colorado River, likely would be shocked by its modern-day taming, and by the complex, hardly equitable distribution formula that greens 1.7 million acres of desert and sends water to 20 million residents in California, Arizona, and Nevada, even as it supplies water to Colorado, Utah, Wyoming, and New Mexico. The Colorado and the other major rivers of the West are so overallocated to farmers and to cities that some have dried up completely. The mighty Rio Grande River that historically sent a steady torrent of freshwater into the Gulf of Mexico now peters out before it reaches the sea. The San Joaquin River no longer flows into San Francisco Bay but rather disappears into a giant plumbing system where it is doled out for agricultural irrigation and drinking water for California's unstoppable growth.[5]

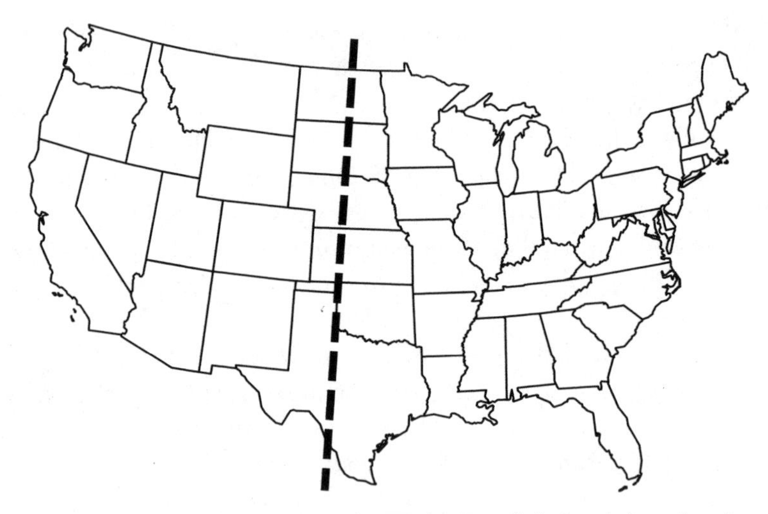

In 1876, federal explorer Major John Wesley Powell declared that a longitudinal line along the 100th meridian divided a moist East from an arid West. Western settlers would have to irrigate and share water. People east of the line had so much water they would never have to worry about it. It turns out they should have. Vast wetland drainage, development, and overuse have led to water scarcity and water wars in many parts of the wet East.
(Map © Christopher Sheek.)

But Major Powell might be even more surprised by the water crisis lapping at the southeastern part of the country. After all, nature graces the South with an average rainfall of fifty inches a year, more than double the amount Powell deemed enough to grow—with no irrigation—any crop that could take the heat.

In the spring of 2004, the nation watched the West with worry as snowmelt in the Colorado, a veritable faucet for cities from Denver to Los Angeles, dropped off by half, resulting in the driest stint in a century of recorded history. By then the man-made backup for western water supply, the major's namesake Lake Powell, had lost 60 percent of its water in a stark reminder that American ingenuity has not quite tamed Mother Nature.[6]

As the western story played out aboveground—the *New York Times* ran dramatic color pictures of Lake Powell's ten-story-high, salt-bleached cliffs—a quieter tale percolated below the soil in the American South. That region, too, was enduring the driest spring in one hundred years. Rainfall deficits of ten inches, near-record-low stream flows, and dried-out soils wreaked havoc on farmers and water managers from Mississippi to Florida.[7]

Traditionally water-rich regions throughout the eastern United States have been threatened in recent years by some combination of overuse and drought. At the National Drought Mitigation Center, housed at the University of Nebraska at Lincoln, director Don Wilhite says researchers who have long worked on the water problems of the West are being called upon increasingly to help cities, farmers, and others in the East.

"We're seeing that the number of basins or watersheds at the point of being overappropriated is increasing. This has long been a problem in the West, and now it's more and more of a problem in the East," Wilhite says. "And we're seeing a tremendous reliance on groundwater in cities in the East as well as the West. Florida is just one of many areas where the groundwater is not going to be able to sustain the growth."[8]

Without the deep reservoirs of the West, many fast-growing eastern cities already were vulnerable to temporary water shortages. Intense population growth and the spread of development have made water problems perpetual. In Raleigh, a combination of overpumping and drought has nearly emptied Falls Lake, the only water supply for North Carolina's capital city. Residents tied up 911 lines trying to report their neighbors for washing cars. Some had to limit showers to four minutes.[9]

In northeastern Massachusetts, parts of the Ipswich River so famous for its namesake clams go completely dry each summer—as soon as the Boston suburbanites who rely on the river for water turn on their sprinklers and fill up their swimming pools.[10] In New Jersey, the Potomac-Raritan-Magothy Aquifer, the state's largest source of drinking water, dipped precipitously as population growth and development hiked groundwater pumping. Water levels dropped a hundred feet, threatening saltwater intrusion.[11]

As water shortages flow east, so do a river of consequences—far more serious than quick showers. In Ipswich, low river flow regularly devastates fish and wildlife habitat, leading to fish kills and closing of the clam beds.[12] During a state of water emergency in New Jersey in 2002, the government halted use of water for construction or use by any new "building, dwelling or structure" in three southern New Jersey townships.[13]

The same year, New York City's water supply reached the most dangerously low levels in more than thirty years, resulting in a drought emergency declaration for the city and four upstate counties. More than 9 million residents were ordered to restrict water.[14]

Today, water managers in a majority of the states believe they will see shortages within a decade, and that is without drought.[15] But nowhere in the country are water shortages more puzzling and prophetic than in notoriously wet Florida—a regular guest on the Weather Channel thanks to its violent hurricanes, thunderstorms, and floods.

NEW STORIES OF SCARCITY

American historians and journalists alike have filled bookshelves with many a good read about water, from Wallace Stegner's biography of John Wesley Powell to Marc Reisner's *Cadillac Desert.* But almost all of those books are set west of the 100th meridian. That made sense over the past century as the West divvied up the Colorado River, built its grand dams and diversions, and lived out dramas from flood to drought.

Today, however, equally theatrical tales of water shortage are playing out in the East. This one is about Florida, but the state is a portrait in miniature of many national and international water problems. In telling water stories about one place, I aim to dig into the bigger questions that surface when people start thinking about water.

One, who owns water? It depends where you live. Western water law is governed by a doctrine called "prior appropriation," which means the one who first dipped his or her straw into the river holds rights to the water, regardless of who owns the land. In the East, so-called riparian rights mean whoever owns waterfront land has the right to "reasonably" use the water.

Where does water come from? In the West, it is mostly surface water, melting off snowcaps and barreling down (or trickling down, as the case may be) the great rivers of the region. In the East, residents rely, too much so, on groundwater, pumped from deep in ancient aquifers.

Like almost every other story about the Sunshine State, this one starts with dramatic population growth. Now the fourth-largest state in the nation, Florida within the decade will balloon another 21 percent, exceeding 21 million people by 2015 to pass New York in population and become the third largest after California and Texas.[16]

Whether Hispanic immigrants or retiring baby boomers, there is one fact the newcomers will not find in the Florida relocation guide. The state no longer has the water to support them all. Officials in four of Florida's five water-management districts report they do not have enough water to supply projected population growth past the year 2025. Groundwater overpumping has led to emergencies in every region. In South Florida, saltwater intrusion has contaminated freshwater supply in some coastal cities. In Central Florida, some neighborhoods have been plagued by sinkholes and dried-up wells. In the Keystone Heights area of North Florida, the problem is bone-dry lakes. Hundreds of families in the region who once lived the waterfront life now look out over docks that stretch uselessly onto white sand and grass.

WATER, WATER EVERYWHERE

Water does not really disappear, of course; the earth is cycling the same water today that the dinosaurs splashed in 100 million years ago. But Floridians have managed to drain, ditch, and divert so much water that there is not enough left in the ground for fast-growing population centers, especially during times of drought. The bizarre tale of how one of the wettest places in the nation managed to get rid of this much water makes up the first part of *Mirage.* Chapter 1 weaves Florida's history of draining swamps and filling wetlands. Chapter 2 reveals America's water patterns, and Florida's remarkably wasteful ones. Chapter 3 reports on the lessons learned as the American West diverted and doled out more water than exists in its rivers, and how eastern states seem headed down the same path. Will the nation's transfer of political power from the Northeast and Midwest to the Sunbelt states result in huge transfers of water, as well? For Great Lakes residents, this is a real and frightening threat.

Chapter 4 is about the often-ironic, sometimes tragic consequences of getting rid of too much water. By drying up their wetlands, for example, Floridians also are drying up their rain. In Louisiana, where coastal wetlands provided a barrier between New Orleans and the sea, Hurricane Katrina proved people need wetlands protection just as much as fish and shorebirds do.

Mirage's middle chapters turn to the politics of water. Chapter 5 explains the greening of Jeb Bush, who led Florida for an unprecedented two Republican terms, from 1999 through 2006. In one way or another, all eight of those years were focused on water: from the massive effort to restore the Everglades, to increasingly severe hurricanes, to the urgent task of finding new water supplies for developers. The chapter tells how, and why, a politician with unhidden disdain for environmental laws and preservation spending during his first, unsuccessful run for governor in 1994 left office a dozen years later with a green reputation. Of course, despite a powerful environmental ethic in Florida, economic development and growth almost always trump protection of natural resources such as water—no matter which political party is in power.

Both politics and water in the Sunshine State once were controlled by agricultural interests, but farmers are quickly losing ground (literal and metaphoric) to home builders, whose influence is the subject of chapter 6. Ever since Florida became a state in 1845, those risk takers who would turn swamps into cities have been able to influence policy in Tallahassee and Washington with their convincing promises of paradise. That is truer than ever today, as Florida's development industry with a $42 billion impact has passed agriculture to become the second-biggest economic driver in the state, behind tourism.

At this writing, Florida's top three political leaders were all developers: Governor Bush a real estate developer, the senate president a home builder, the house Speaker a paving contractor. Not only are Florida's home builders among the largest campaign contributors, to both Republicans and Democrats, at both state and national levels, they are also key political advisers. While CEO of WCI, Florida's largest home-building company, for example, Al Hoffman was also finance chairman for the Republican National Committee during both of George W. Bush's presidential campaigns.

This is how, unbeknownst to the average U.S. taxpayer, Florida's water woes now command attention, as well as billions of federal dollars, at the national level. While citizens across the nation might agree with spending $10 billion to save America's Everglades for environmental reasons, the elected officials who don khaki garb for photo ops in the swamp have an equally important motive: urban water supply. The Comprehensive Everglades Restoration Plan now underway in South Florida is the largest public works project in the history of the world. While sold to the public as an environmental project, it is equally a strategy to increase water supply to the cities of southeast Florida.

Mirage's latter chapters explore the conflicts and controversies inevitable when a resource becomes scarce. Foremost are the water wars, the subject of chapter 7. Florida, Alabama, and Georgia have fought for more than twelve years over the Chattahoochee and Flint rivers, which supply water to metro Atlanta on their way to Florida's Apalachicola Bay. And politically powerful South Florida has challenged the bucolic but politically weak northern half of the state for its historic Suwannee River as yet another water source to quench future growth. Many other water-rich areas of the eastern United States, including the Carolinas, also are fighting over water resources due to intense growth and unclear rules about who is entitled to them.[17] The Great Lakes region, meanwhile, is gearing up for the day its leaders think is inevitable: when thirsty parts of the United States come looking for water.

Chapter 8 tackles the bottled-water industry. Why do bottlers in Florida and around the nation get to pump groundwater for free, then sell it at an eye-popping profit, just as citizens are being asked to wean themselves off groundwater and spend money on costly alternatives such as desalination plants?

Chapter 9 explains the weird economics of water. Why is water so cheap? More mysteriously, why do those of us who are well-off pay so much less for water than those who are poor? This is a truism globally and locally. In Palm Beach County, Florida, residents of the exclusive east coast island of Palm Beach use, on the average, more than four times the water but pay half as much for it than the farmworkers of Belle Glade on the west side of the same county.

Chapter 10 reveals how private companies in Florida and around the globe are posturing to profit from increasing water shortages. Speculators such as Texas corporate raider Boone Pickens are gearing up for the time when water, from Texas on east, will be bought and sold like pork bellies as it is in much of the West.

Chapter 11 confronts the concerns and the promise of new technology. In addition to its massive Everglades restoration project, Florida plans an orgy of high-tech water-supply initiatives, from desalination plants to aquifer wells that pump water during wet times and store it deep underground to pull up in periods of drought. If the state's leaders have their way, federal taxpayers will help them fund part of this effort, too. Florida's congressional delegation is pushing federal desalination legislation to help fast-growing areas pay for the expensive plants.

In the twentieth century, all Americans footed the bill for the huge water projects that subsidized development of the arid West. In the twenty-first, taxpayers are being asked to help fund costly new water supplies for the significant population shift now underway to the nation's Sunbelt. Before they fork out these billions, Americans may want to ask a few questions: Will eastern states take heed of mistakes made in the West, where many waterworks have devastating ecological consequences, and where most rivers are so overallocated that during times of drought there is not enough water for legal users—no less for fish and wildlife? And, faced with a new specter of water scarcity, will flagrantly wasteful states such as Florida learn from conservation-minded places like California that continued growth and development do not have to mean higher and higher rates of water consumption?

STATE OF MIND

The remarkably wasteful ways we use water, the legal and natural workings of water, its vibrant social and political history and puzzling economics: all demand understanding as more and more of America faces water scarcity.

A century ago, Floridians thought their biggest problem was too much water where people wanted to settle. Now, our biggest problem is that we do not have enough water where people want to settle. How did this about-face occur in just one hundred years?

Digging for the answers starts in the muck of Florida's once-too-wet lands, in the moxie of its settlers. Those intrepid souls got rid of water at every turn, draining wetlands by the thousands of acres and filling them in.

Just as important as the physical remaking of water and land in Florida was the remaking of the state in the American imagination. When Florida was under water, its pitchmen divvied it up and sold it by the acre, hawking mosquito-infested swamps as tropical paradise. Today, even as water-related headaches such as sinkholes plague families like the Atteberrys, a new generation of boosters works to spin a myth of Florida as oasis.

Like a family suspending its reality during a Disney World vacation, being a Floridian means buying into the myth. Just ask David and Vivian Atteberry. The hole that opened in their middle-class subdivision sank more than their custom-built home. During a fight with their insurance company, the Atteberrys lost their house in foreclosure. The ordeal forced them back to Illinois, where, in their fifties, they now live with Mrs. Atteberry's elderly mother.

But in their minds, the Sunshine State is still a dreamy paradise. They say they will return, if they ever get the chance. "We left because we didn't have a place to live and we were desperate," Mr. Atteberry says. "But we didn't want to leave. We love Florida. If we could, we'd be there today."

I

History & Myth

LA FLORIDA. Spanish explorer Juan Ponce de León gave the peninsula its modern name when he stepped ashore on the Atlantic coast in 1513. It was Easter season, known in Spanish as the *pascua florida.*

You have no doubt heard that when he discovered Florida, Ponce de León was searching for the Fountain of Youth. One sip from the magical fountain would cure any ailment, keep a man young. To this day, schoolchildren learn the story in fourth grade. But it is not true. The best historical sources, including the highly detailed charter given to him by the king of Spain, offer no evidence that Ponce de León ever gave thought to a Fountain of Youth, much less that he sailed to what is now Florida to look for it.

The truth is that Ponce de León carried a contract to settle the lands he discovered and to enslave native people. Such is myth. If it is more appealing than the real story, and if it endures long enough, it can overcome truth.

Florida has always drawn powerful mythmakers, so bold in their visions that their made-up stories have often come true.

Consider: On the eve of the twentieth century, Florida was home to only 529,000 souls, and among the poorest states in the American South.[1] Even as land speculators pitched the place as a balmy, beautiful land of opportunity, they mourned privately over worthless swampland and malarial living conditions. The most famous, a Philadelphia industrialist named Hamilton Disston, set about to drain the Everglades. Instead, the great swamp drained him. He committed suicide in his bathtub in 1896.

Fast-forward one hundred years. Florida met the twenty-first century as one of the most influential states in the nation: the fourth most populous, it would soon overtake New York as the third behind California and Texas. Floridians decided the 2000 U.S. presidential election. South Florida was a bridge to Latin America, with Miami the region's headquarters for banking and commerce. Orlando was the top destination for tourists in America. And Floridians had jobs: as the nation's jobless rate rose in the early years of the century, Florida's remained among the lowest of all the states.

From backwater to bellwether in one century. How? Much credit is due John Gorrie, the Florida physician who pioneered air-conditioning. But those most responsible for Florida's remarkable turnaround were blowing not cold but hot air. More than anything, the state's destiny was shaped by a colorful succession of pitchmen—and some noteworthy pitchwomen. Some, like the social reformer Harriett Beecher Stowe, promoted the "raw" and "unsettled" state in an earnest effort to save it.[2] Others, like Disston, made outrageous claims to pump up land sales. "You secure a home in the garden spot of the country, in an equable and lovely climate," boasted one of his ads, "where merely to live is a pleasure, a luxury heretofore accessible only to millionaires."[3]

The plot may well have been under water.

ENDLESS LAND, ENDLESS WATER

When Florida entered the Union in 1845, it was an "empty, often impenetrable and endless land" with 57,951 residents.[4] Their first state flag said this: "Let Us Alone." Statehood was supposed to bring settlers. But with the western frontier still wooing adventurers, Florida was a hard sell—particularly after January 1848, when California miners found gold at their camp on the American River at Coloma near Sacramento. By 1860,

Florida's population had reached only 140,423: 77,746 whites, 61,745 slaves, and 932 free blacks. Most of them were clustered in a thin band along the northern border of the state.[5]

Their part of Florida was a deep-wooded wilderness, much of it dark and wild. Tall longleaf-pine forests dominated the uplands and coastal plains; hardwood hammocks thick with granddaddy oak, beech, and magnolias spread across the river basins. Black bears and panthers tamped down pine-needle paths through thick saw palmetto. On sunny days, so many alligators clambered out of rivers to dry out and sleep that the banks looked paved with black granite.

More than anything, the region was defined by water. In these days long before the Intracoastal Waterway, salt marshes stretched for miles inland from the Atlantic and Gulf coasts. In the interior, three grand and powerful rivers—the St. Johns in the east, the Suwannee in the center, the Apalachicola in the west—quartered the state. They flowed in and out of thousands of acres of cypress swamps, ponds, and lakes of every imaginable size. Throughout those waters bubbled up hundreds of ice-blue springs from deep below ground, reminders of the constant and forceful flow of water through underground channels in Florida's limestone core.

Exploring the St. Johns River in 1774, the naturalist William Bartram described one such spring as an "enchanting and amazing crystal fountain, which incessantly threw up, from dark, rocky caverns below, tons of water every minute, forming a bason, capacious enough for large shallops to ride in, and a creek four or five feet depth of water, and nearly twenty yards over, which meanders six miles through green meadows, pouring its limpid waters into the great Lake George, where they seem to remain pure and unmixed."[6]

But to most pioneers, Florida was not so enchanting. Across the nation's frontier, nineteenth-century homesteaders viewed the wilderness as maleficent, their moral purpose to destroy it.[7] President Andrew Jackson asked in his 1830 inaugural address: "What good man would prefer a country covered with forests and ranged by a few thousand savages to our extensive Republic, studded with cities, towns, and prosperous farms, embellished with all the improvements which art can devise or industry can execute?"[8]

In Florida, the conventional wisdom was that the peninsula was far too wet, a malady that came with frightening symptoms, from poisonous snakes to cloudlike swarms of mosquitoes. Wrote one new resident in 1872: "From what I have observed, I should think Florida was nine-tenths water, and the other tenth swamp."[9]

Indeed, until the late 1800s, most of South Florida was a vast expanse of interconnected wetlands that covered 8.9 million acres and extended from the lakes and marshes south of today's Orlando area all the way to the bottom tip of the state. In this region, too, a mysterious marsh flowed for a hundred miles from Lake Okeechobee to Florida Bay. An 1846 U.S. survey map called it "Pah-Hay-O-Kee or Grass Water, known as The Everglades."[10] The marsh was a jungle of wildlife—the only place in the world home to both crocodiles and alligators. It became the last major sanctuary in the eastern United States for egrets and herons, threatened with extinction during the Gilded Age because of a fashion craze for plume-decorated hats. During nesting season, the southern tip of the marsh was so thick with plume birds that it looked like "a white cloud, a cloud in constant motion."[11]

Also hiding out in the marsh was a small number of Seminole Indians who had fled their woodland homes during the United States' three tragic wars against the tribe. In fact, the bloody accounts of Indian wars in these wetlands first piqued the federal government's interest in the Everglades, writes Glades historian David McCally.[12]

In 1847, the government sent a respected St. Augustine lawyer named Buckingham Smith to investigate the marsh, particularly whether it could be "reclaimed and made valuable." His lyrical report described "a vast lake of fresh water extending in every direction from shore to shore beyond the reach of human vision, ordinarily unruffled by a ripple on its surface, studded with thousands of islands of various sizes, from one-fourth of an acre to hundreds of acres in area, and which are generally covered with dense thickets of shrubbery and vines."[13]

"A solitary inducement can not now be offered to a decent white man to settle in the interior of the Everglades," wrote Smith, who proclaimed the region suitable only as "the haunt of noxious vermin, or the resort of pestilent reptiles."[14]

Even Floridians who, like Smith, saw beauty in the swamps nonetheless dreamed of draining them dry. Alexis de Tocqueville pondered this on his American travels, during which frontierspeople considered him insane to want to experience wilderness for pleasure. Tocqueville figured Europeans yearned for nature because they had lost it, while having to live in the wilds made Americans prize "the works of man" over those of nature. "Their eyes are fixed upon another sight," Tocqueville wrote in *Democracy in America.* They "march across these wilds, draining swamps, turning the course of rivers, peopling solitudes, and subduing nature."[15]

Buckingham Smith believed draining the Everglades was not only feasible but that it would reveal a fertile expanse of soil suitable for plantations that would rival the West Indies for tropical fruits, coffee, tobacco, and sugar. He estimated it would take $500,000 and less than a decade.[16] Eager politicians and adventuresome businessmen latched on to his vision. They would not let go of it for the next century.

In 1845, the United States had welcomed its twenty-seventh state with a gift of 500,000 acres of federal land, much of it submerged, for "internal improvements." In 1850, Congress passed the Swamp and Overflowed Lands Act, which turned over any wetlands considered unfit for cultivation to the states to drain for settlement. Florida's take was more than 20 million acres. The young state's elected officials spent five years fighting over how to manage their newfound land wealth. Finally, in 1855, they came up with the Internal Improvement Fund, with a board of trustees composed of the governor, comptroller, treasurer, attorney general, and registrar of state lands.[17]

These board members, along with other state leaders, set upon the path their successors would follow for the next 150 years: giving natural resources away to foster private development. The legislature passed the Riparian Rights Act in 1856, handing over state-owned sovereign lands to shoreline owners who would construct docks, wharves, and other projects to spark commerce and growth. The Florida Supreme Court later ruled this handout violated the U.S. Public Trust Doctrine, which obligates states to hold public lands and waters in trust for the benefit of all residents.[18]

Florida joined the Confederacy; some believe the Confederate treasury is buried outside Gainesville. During the Civil War and long after, the state remained sparsely populated and extremely poor—with little going for it save the land and the water. Desperate to grow, the trustees dangled these two primary assets to would-be developers, often handing them over for nothing. The water and land giveaways often proved foolhardy. Yet Florida's leaders continued them into the twenty-first century.

In the twentieth century, the big freebie was groundwater. Back in the nineteenth, it was swamplands. In one particularly disastrous decision by the Internal Improvement Fund, antebellum railroad and canal companies that had forged track and ditches before the Civil War convinced the trustees to put up the state's swamplands as collateral for their bonds. This violated the Swamp and Overflowed Lands Act requirement that the wetlands be drained for human settlement. But the railroad men argued their work would open the wild interior to homesteaders. Instead, devaluation of the Confederate currency plunged them into bankruptcy, and they defaulted on their loans. Bondholders rushed down to Florida to claim millions of acres of land. Court injunctions prohibited Florida from using its lands until the creditors were paid.[19]

But Florida was broke and faced losing millions of acres worth far more than the debt. Trustees and state lawmakers tried for more than a decade to squirm their way out of paying. In fact, the Florida legislature kept trying to hand over the land it no longer controlled to developers. In 1879, lawmakers passed a bill ordering the trustees to award ten thousand acres of swampland for each new mile of canal or railroad track built. Again, creditors got court injunctions to stop the giveaways.[20]

Finally, in 1881, newly elected governor William Bloxham (1881–85) struck an ambitious economic development gambit with Hamilton Disston to bail Florida out of the crisis.[21] In a real estate deal the *New York Times* called the largest land purchase by an individual in the world, Bloxham arranged to sell Disston 4 million acres for 25 cents each to pay off the creditors and free the land from the courts.[22] In return, Disston and his partners would drain 12 million acres of swampland to the north and west of Lake Okeechobee in exchange for half of what they drained, or 6 million acres.[23]

The deal handed Disston control of nearly half of Florida—including the still-uncharted Everglades. Disston promised "to drain all the lands overflowed by Lake Okeechobee, the Kissimmee River and its branches, and contiguous lakes . . . to lower the level of Lake Okeechobee, to deepen and straighten the channel of the Kissimmee River, and to cut canals and ditches to connect Lake Okeechobee with the Caloosahatchee River on the west, the St. Lucie River on the east, and the Miami and other rivers in the southeast."[24] Mark Derr, a chronicler of humans and the land in Florida, notes that "the final configuration of south Florida's maze of canals, dikes, and channelized rivers generally follows this blueprint, but it took nearly a century and more than $500 million to accomplish."[25]

Not only in the Everglades, but across the peninsula, Floridians were in a frenzy to get rid of the water. The state's first water laws were the so-called ditch-and-drain laws of 1893. They authorized counties to "build drains, ditches or water courses upon petition of two or more landowners."[26] No one could have imagined one hundred years later developers would petition state agencies to try to find water for growth.

While Disston went to work draining the swamps, Governor Bloxham paid off the creditors and opened up millions of Florida acres to railroad construction. When Bloxham took office in 1881, Florida had 550 miles of rails. By 1891 there were 2,566 miles; by 1900 there were 3,500.[27] The lines included Henry Morrison Flagler's East Coast Railroad, stretching from Jacksonville to Miami, and Henry B. Plant's lines along the west coast.

Flagler, Plant, and others tried to outdo each other's promotions, publishing guidebooks that advertised their railroads and grand hotels. Great Atlantic Coastline Railroad hired the Southern poet Sidney Lanier to write a guidebook called *Florida: Its Scenery, Climate and History* (with a chapter for consumptives). Lanier started his book with a note on the genre: "So much being said about the abundant protection of strict truth, one can now go on to detail (without the haunting fear of being classed among the designing hysterical ones) the thousand charms of air, water, tree and flower which are to be found in Florida, and which remain there practicable all the winter days."[28]

Florida's most famous promoter was Harriett Beecher Stowe, who wintered on the St. Johns River. The wildly popular author of *Uncle Tom's Cabin* cranked out articles about Florida life for the religious press and published her ode to the state, *Palmetto Leaves,* in 1873. Stowe well knew Florida's "roughness," "hardships," and "loneliness." Still, she recommended it to three types of people: those suffering from pulmonary or rheumatic conditions, who might find Florida "the salvation of life"; those who enjoyed nature and outdoor activities like fishing and hunting; and the "industrious young man" who might start off planting citrus trees and within a decade "realize a handsome independence."[29]

The late water historian Nelson Manfred Blake observed that Stowe's enthusiasm made a strong impression on Northern readers. If a personality "denounced as a disturber of the peace in the antebellum South could find a happy second home in Florida, other Yankees could also expect to be well treated here," Blake wrote. "And more and more of them visited the state each winter."[30]

During the depression of the 1890s, Hamilton Disston's income from land sales lagged far behind the huge expenses he had incurred trying to drain the Everglades. He began to mortgage his Florida holdings for loans and more loans. His situation became increasingly desperate after the Panic of 1893. No longer able to pay his workers, he had to halt his dredging operations. Banks called in his loans, and he defaulted on bonds. On the night of April 30, 1896, after attending the theater in Philadelphia, he filled his tub with water, climbed in, and fired a bullet into his head.[31]

Now, these new bondholders joined hundreds of other claimants Florida had to fight over the legislature's overly generous land grants. As the state's politicians sobered up from their decades-long railroad and canal-building binge, they realized that they had no more land to sell or grant. They had promised more land than the state even owned.[32]

The Populist governor Napoleon Bonaparte Broward (1905–9) won election by painting the railroad companies as land thieves, and promising to drain the Everglades once and for all. More and more political and business leaders had begun to deride the idea as "a great scheme to drain the State Treasury. . . . The attempt will simply result in a waste of the people's money."[33] But Broward traveled up and down Florida to convince voters that state-funded drainage was a crucial investment in Florida's economic future.

Real estate developers certainly thought so. Those who specialized in swampland sales put enormous pressure on Florida's politicians to keep the state dredges working to get rid of water. Only if reclamation was underway could they sign up new customers and keep the old ones paying their monthly installments. It is sort of a mirror image of today: in the twenty-first century, Florida's home builders put enormous pressure on state and local politicians to *find* them water so they can break ground on new developments and presell homes and condos.

While buyers who invested in the Everglades muck hoped for a long-term payoff that seemed less and less likely, those lucky enough to have found Florida's palm-shaded southeast coast got in on the most spectac-

The first dredge in the Florida Everglades, circa 1907.
(Courtesy of the State Library and Archives of Florida.)

ular real estate boom in American history. At least while it lasted. Increasingly accessible by train, and then car, the tropical cities around Miami drew Florida's first population boom—2.5 million people in 1925 alone. That year, mainland property six and eight miles outside Miami was selling for $26,000 an acre. Just a decade before, pitchman Carl Fisher had offered up Miami Beach property for free to anyone who would come down and settle it. This torrent "exceeded anything the nation had seen in oil booms or free-land stampedes," wrote the Florida historian Michael Gannon.[34]

To keep up with the boom, developers opened one subdivision after another of hastily built homes on hastily built land. In Fort Lauderdale, for example, Charles Green Rodes invented "finger-islanding" to sell more high-priced waterfront lots: he dredged a series of canals at right angles to the New River, pulling out mangroves as he went, and used the fill to make new land. Down each artificial finger he built a street, dead-ended at the river for privacy. Each home was waterfront, with easy access to the river and the entire east coast canal system. The success of this "dredge-and-fill" development would make it a favorite of Florida home builders for the next forty years.[35]

Scarcely heard over the squawk of the real estate salesmen who called from bullhorns along Miami Beach, some Floridians decried the loss of the environmental treasures that made the place so inviting in the first place. Flagler had struck nature a severe blow before the turn of the century by dredging Biscayne Bay and the Miami River, obliterating their deep blue springs and thick vegetation.[36] In 1923, the naturalist Charles Torrey Simpson asked: "What natural beauty will we have left for another generation? What right have we to waste and destroy everything nature has lavishly bestowed on the earth?

"The only attraction belonging to the state that we do not ruin is the climate, and if it were possible to can and export it we would do so until Florida would be as bleak and desolate as Labrador," Simpson wrote in his *Out of Doors in Florida*.[37]

He was wrong, of course. People were ruining the climate, too.

Inevitably, Florida's first real estate bubble would burst. The boom came to an end in January 1926, with failed banks, ruined speculators, and northerners cheated out of their savings. That fall, a great hurricane would further shatter the dreams of Florida's boosters. The September storm hit the southeast coast with 130- to 150-mile-an-hour winds and moved across the interior, killing some 372 people, injuring 6,000, destroying 5,000 homes, and leaving 18,000 without shelter. Just two years later, in 1928, another September hurricane, the deadliest in state history, would roar through southeast Florida, taking more than 2,500 lives.[38]

Florida's hopes of reclaiming the Everglades had already dimmed, as decades of land giveaways, overbonding, and swampland sales by governors and the Internal Improvement Fund trustees caught up with the state, leaving the project hopelessly in debt.[39] The deadly hurricanes inspired a new partnership between the state and federal government, and a fundamental change in the way they managed South Florida's water. Drainage would no longer be the order of the day. Flood control would take its place. Of Florida's few remaining Everglades holdings, 1.3 million acres were set aside for water conservation; another 800,000 acres south of Lake Okeechobee designated agricultural.[40]

The real estate collapse and bank failures in 1926 plunged Florida into the Great Depression three years ahead of the rest of the nation.[41] The Depression era saw in Florida the one and only instance in the state's history in which leaders and promoters made any serious effort to slow growth—at least growth of a certain sort. For three winter seasons in the 1930s, state police took the extreme step of patrolling highway entrances to Florida and turning back travelers who lacked provable means of support, many of them desperate families pouring south in search of work and warmth. In the winter of 1935–36 alone, an estimated 50,000 people were turned back at the state line.[42]

Still, Florida's boosters kept spinning their sunny myths of prosperity and opportunity. Within a decade, their stories would begin to come true.

POPULATION BOOM

By December 1941, Florida was seeing signs of economic recovery from the Depression—and growth. Along the east coast, Highway A1A carried a steady stream of two-way traffic buzzing past mom-and-pop motels, seaside trailer parks, and a new retail phenomenon: roadside citrus shippers where visitors could grab a sack of Florida navels or send one north.

The Japanese attack on Pearl Harbor stopped tourism in Florida as fast as lightning clears a swimming pool. But U.S. entry into World War II brought another stream of visitors to the Sunshine State. They were not wearing bathing suits. In the words of Florida historian Gary Mormino, these newcomers came in khaki, olive drab, and navy blue.[43]

At the start of the war, America's generals had a problem: the country's military training facilities were overburdened. When they looked around for a place to house hundreds of thousands of soldiers, they found it in Florida—and its empty resorts. Soon, the state's hotels were barracks, its restaurants mess halls, its white-sand beaches vast training grounds with formations of soldiers sweating in the tropical heat.

In all, the state was home to 172 military installations, including the army's basic-training center at Camp Blanding, during the war Florida's fourth-largest city after Jacksonville, Miami, and Tampa. (Two thousand German prisoners, some captured in U-boats prowling Florida's coastline, were jailed there as well.) The federal government poured hundreds of millions of dollars worth of war contracts into the state. And by the end of the war, more than 2.1 million men and women had called Florida home for military training.[44]

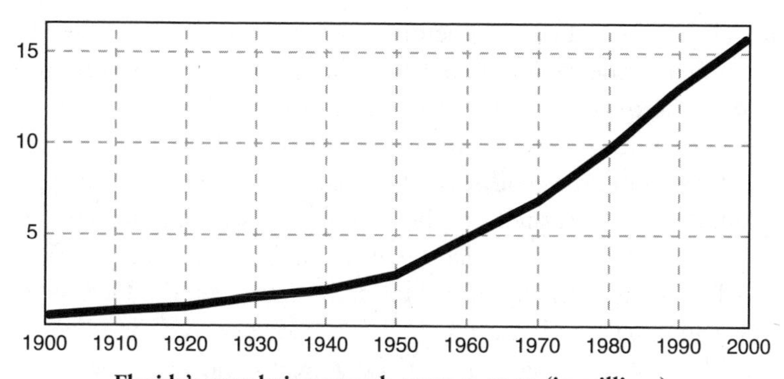

Florida's population growth, 1900 to 2000 (in millions).
(Data from U.S. Census Bureau. © Christopher Sheek.)

This exposure would be key to the dramatic population growth of the postwar era. It was the introduction of the American GI to Florida—crucially combined with the growing uses of air-conditioning and mosquito control—that made the state accessible to the middle class.[45] After the war, families began to pour into the peninsula, not just to vacation anymore, but to set up households, make a new start. Florida exploded with new residents between 1945 and 1960, with population growth averaging 558 people a day during the fifteen-year period.[46]

Finally—growth! Florida's leaders were euphoric. They looked forward to a surging economy with lots of jobs. Despite the massive migration and new residents' complaints about poor schools and inadequate services, they continued to beat the drum for more. "Emphasizing low taxes, a healthy environment, inexpensive land, and a pro-business political climate, state agencies, led by the governor's office, appealed to potential investors throughout the nation," write the Florida scholars David Colburn and Lance deHaven-Smith.[47]

By the 1950s, it would have been hard to believe that sandy beaches were natural to only a few parts of Florida—on Cape Sable and on the ocean side of some barrier islands. The entire remainder of the coast, once thick with red, black, and white mangroves, labyrinths of roots and trunks that made the perfect nursery for young marine animals, was being chewed up by the mile.[48]

The assault on Florida's environment that had been thwarted by the Depression now grew more frenzied every day. The top priority was the same as it had been a century before: getting rid of the water. Wetlands were drained by the thousands of acres. Along the coastlines, hundreds of miles of mangroves were bulldozed and left to rot. Huge sand dunes that graced both sides of the peninsula were flattened and used for fill—replaced with buildings and seawalls. Back in business after the lean war years, developers returned to the old finger-islanding techniques to build, the old installment plans to sell. In just one small scene that repeated over and over in South Florida in the 1960s, the Deltona Corporation in 1964 divided up a deep-forested, twenty-four-square-mile island called Marco, a sanctuary of tidal creeks and mangrove swamps teeming with fish and wildlife—a nursery for redfish and drum, a rookery for roseate spoonbills, snowy egrets, wood ibis, and countless other birds. The company sold lots to finance a dredge-and-fill operation that would chop the island into a stark, treeless geometry of roads and canals to create 5,700 waterfront lots.[49]

Charles Green Rodes's dredge-and-fill technique had become particularly deadly to Florida's ecosystems, once dominated, of course, by wetlands. In 1957, lawmakers in the state capital, Tallahassee, tried to gain some control with a law called the Bulkhead Act. Local governments now had to issue dredge-and-fill permits and establish "bulkhead lines" to define the outer limits of waterfront development. The permits had to be okayed by the trustees of the old Internal Improvement Fund. But environmentally destructive filling remained rampant. One county, Pasco, set its bulkhead line two miles out into the Gulf of Mexico.[50]

Luther J. Carter, one of the first journalists to chronicle Florida's environmental losses, reported in 1974 that "more than 60,000 acres of prime estuarine habitat for fish and shellfish had been lost to dredging and filling alone."[51]

Just as Florida's early boosters paid too dear a price for development in the days of the land giveaways, those at midcentury could not seem to see what they were sacrificing for growth. They also made a key miscalculation. They figured state and local governments would become so rich with tax revenues from the new residents that there would be plenty of money to fund their needs: new schools, new highways, much-needed sewage-treatment plants. They were wrong. By 1960, the pace of development had far outrun that of infrastructure. Pollution began to plague Florida's rivers, lakes, and canals. Only one sewage-treatment plant had been built in the entire state. For the most part, raw sewage was piped straight into rivers, the Atlantic, or the Gulf. A series of pipes off Miami Beach spewed millions of gallons of untreated sewage each day into a site visible from shore. It was dubbed the Rose Bowl for its odd pink tinge.

Then as now, Florida was a schizophrenic place, at once touting and destroying everything that made it feel like paradise—the granddaddy sand dunes, the unobstructed ocean views, the fat redfish and long-snouted snook that hid out in mangrove bushes.

Even as the concrete poured, Florida's magnificent natural areas and wildlife had kept the state at the vanguard of the conservation movement. In 1901, for example, Florida became among the first states to outlaw plume hunting. In 1903, U.S. President Theodore Roosevelt established the first U.S. wildlife refuge in Florida—at Pelican Island in the Indian River Lagoon. And Florida's eminent environmentalist Marjory Stoneman Douglas wrote her famed *The Everglades: River of Grass* in 1947—more than a decade before the dawn of the popular environmental movement in the early 1960s. *River of Grass* "reshaped the way America saw the southern tip of its southernmost state,"[52] giving Americans, for the first time, a view of the Everglades as a national treasure rather than a dismal swamp.

But destruction of wetlands, pollution, and other problems associated with growth were outrunning conservation efforts. In 1962, Rachel Carson published *Silent Spring,* the book that opened ordinary Americans' eyes to the threat of toxic chemicals and explained humans' ecological dependence on the earth. In Florida, that was the year bald eagles reached their lowest point, due to both pollution and habitat degradation, with an estimated 251 nests statewide.[53] As the future governor Bob Graham—a developer himself—was fond of saying, Florida had become so attractive that "the walls of constraint collapsed under the desire to find a home in the sun."[54]

"WATER WONDERLAND"

No one represents Florida's mid-twentieth-century developer as perfectly as a clean-scrubbed, handsome pair of brothers: Leonard and Jack Rosen. And no place exemplifies how tall tales come true in Florida better than the city the Rosens built on a point jutting out from the state's southwest coast: Cape Coral, nicknamed Florida's Water Wonderland.

The Rosens were born to Russian-Jewish émigrés who had settled in

Baltimore, Maryland. They honed their remarkable sales skills as teens, after their father was run over by a streetcar and killed during the Depression. In the 1930s they hawked coat hangers and other household goods at carnivals and on the boardwalks of Atlantic City. During the 1940s they built a furniture empire by selling to families on installment. By the 1950s they had tapped television, coming up with long-form, storytelling commercials to sell lanolin hair products.[55]

The brothers used all those tricks and more to hard-sell a city that existed only in their dreams. In 1957, they bought 1,724 acres of palmetto scrub near Fort Myers for a $125,000 down payment. Ten years later, they were multimillionaires at the head of the largest development corporation in the country, Gulf American Corporation, bringing 50,000 visitors a month to Florida to look at yet-to-be-built Cape Coral. (The Rosens bought into the marketing superstition that developments had to have repeating initials, thus their Florida legacies Cape Coral, Golden Gate, River Ranch, Remuda Ranch, Barefoot Bay, and Rio Rico.)[56]

The Rosen brothers convinced thousands of small, installment buyers from throughout the United States and Europe that this mangrove-choked, mosquito-infested scrubland where the Caloosahatchee River empties into the Gulf of Mexico, with only one dirt road in and out, would someday be a paradise of waterfront homes, lush tropical landscaping, and country-club amenities.

The pitch was part scam: Gulf American was notorious for selling land it did not own, for switching homesites after sales were made, and for making it almost impossible for small owners to resell their lots. But there is no denying that most of Leonard and Jack Rosen's promises did come true—beyond anyone's wildest dreams. Fifty years ago, Cape Coral did not exist. Today, it is the fastest-growing city among those with populations over 100,000 in the eastern United States. It has the waterfront homes (thanks to some 400 miles of canals), the palm-lined streets, the yacht club, the sparkling pools, the green golf courses.

All this comes with a steep environmental bill. Today's Cape Coral residents are still paying it. The moving of so much land and water, combined with overdevelopment and drought, has resulted in a water-supply crisis in the Water Wonderland. Today, water levels in Cape Coral are so low that 4,000 homeowners are watching their wells go dry. Meanwhile fire department officials, who depend on the city's canal system for hydrants, say about a third of Cape Coral is at risk of not having enough water pressure in case of a major fire.[57]

To build up Cape Coral, Gulf American ignored Florida's new dredge-and-fill laws and scooped millions of cubic yards of fill from the Caloosahatchee River without permits. The company obliterated thousands of acres of mangrove habitat. Said one wildlife biologist in 1973: "They've developed a Sahara Desert down there. Every time the wind blows, the dust flies. They've completely denuded the land of every bit of vegetation. It's a bad one—an abortion."[58]

Such practices were standard operating procedure for the times. Florida's local governments were not equipped to stop them. Through the 1960s, Lee County, home of Cape Coral and several other Gulf American communities, had no master plan and virtually no zoning. In fact, by 1970, more than half of Florida's counties still had no planning or zoning whatsoever.[59]

But amid the burgeoning environmental consciousness of the 1960s, Florida would become among the first states in the nation to take bold steps to protect water, land, and wildlife. The man who led the way was an unlikely green: a one-term governor named Claude Roy Kirk Jr. with no personal interest in conservation. The charismatic but quirky former marine had been elected in 1966 in a stunning victory that made him the first Republican governor of Florida in the twentieth century.[60] He became the first chief executive in state history, and the first in the South, to make the environment a centerpiece of his governorship.

Kirk's inspiration was Nathaniel Reed of Hobe Sound, the governor's dollar-a-year environmental adviser. Reed, reared in a wealthy Connecticut family that spent half the year in the then-wild tropical paradise of Jupiter Island, was so rabid a nature lover that his mother said he "came out of the womb casting a fly rod."[61]

Reed convinced Kirk to champion Florida's environment, a repayment to Mother Nature that he would soon repeat at the national level. In 1971, President Richard Nixon—another politician with little interest in nature—asked Reed to serve as assistant secretary of interior and preside over the greatest handout of conservation monies in U.S. history under the Land and Water Conservation Fund. Reed helped convince both Nixon and Kirk of the extraordinary political capital to be tapped in the nation's growing environmental ethic. Opportunistically, both men built unprecedented environmental records, with President Nixon creating the Environmental Protection Agency (EPA) by executive order, and signing into law the Clean Water Act, the Clean Air Act, and the Endangered Species Act. (Nixon vetoed the Clean Water Act of 1972 on cost grounds, knowing that he would be overridden by Congress.)[62]

Governor Kirk's delight in ruffling the feathers of Florida's power brokers, combined with his love of a press-generating cause, made the environmental platform an easy sell. "He had an interest in the lowest guy on the totem pole and at that time, conservation was the lowest thing in anybody's mind in Florida," remembered Reed. "It was rape and run, avarice and greed. Make money now, and do not worry about the future."[63]

At Reed's urging, Governor Kirk championed two key conservation measures his first year in office. One was the Florida Air and Water Pollution Control Act of 1967, which created a board and permitting process to oversee air and water pollution discharges. The other was an amendment to the 1957 Bulkhead Act. The 1967 update required a biological survey as a prerequisite to any state or local decision allowing alterations to tidal lands or the bottoms of state-owned lakes.[64]

Known as the Randell Act, the new restrictions to dredge and fill were named for their sponsor, state representative M. T. "Ted" Randell of Fort Myers, who had become alarmed by impacts on his favorite fishing holes. It directed the cabinet to weigh the public benefits of proposed waterfront development against environmental losses. From that point on, owners of submerged lands in Florida could no longer assume they could be filled or used as fill. Also at Reed's urging, Governor Kirk frequently opposed dredge-and-fill permit applications, which made it difficult for the six Democrats who made up the cabinet to approve them.[65] In the wake of the Randell Act, approvals for the permits, which had been averaging two thousand a year, dropped to two hundred a year in the last three years of Kirk's term.[66]

Despite the early rolls of an environmental drumbeat, however, Kirk and other Florida leaders continued to march with the growth band. Kirk's appetite was whetted particularly by a man named Walt Disney, who claimed his idea for a "City of Tomorrow" could bring as many as 13 million visitors a year to the Sunshine State. Kirk greased the project mightily. On one famous occasion, he flew to California in the middle of a teachers' strike crisis back home to visit Walt Disney's brother, Roy. On that day, Kirk said, he assured Roy Disney "that our effort would be to be sure that no politician, no city, no code enforcement people would put their arm on them either legally or illegally to make them change the nature of what they were trying to do."[67]

Walter Elias Disney had acquired his 27,500 acres in Central Florida in 1964 for an average of $200 an acre. It helped that they were mostly underwater. The dreamer who had spent his boyhood doodling whimsical pictures instead of studying, who grew up to create the character of Mickey Mouse and hundreds of animated cartoons, envisioned an ultramodern fantasy city on those acres where others could see only swamp.

"No, no, this won't do," Walt Disney said the first time he saw the property's water—stained black by tannic acid from the cypress trees that dominated the land.[68] He wanted the wetlands drained. He wanted the trees gone, replaced with civilized flora and fauna. Most of all, he wanted the water blue.

He got all that, and much more. In a handover of power to private industry that was astonishing even by the standards of the state's long-boosterish governors and lawmakers, the Florida legislature in 1967 gave Disney governmental power over its own land. Walt Disney had implied the legislation would make or break his grand move into Florida. The package of three bills turned the company's holdings into two cities and the Reedy Creek Improvement District. Since no one lived there but a few Disney employees—that is still true today—the move handed the company all the rights and responsibilities that would have fallen to the two counties its acreage spanned: Orange and Osceola.

The legislation passed just months after Walt Disney died of lung cancer—he was an aficionado of tiny cigars. It ensured his project could remain "in a state of becoming," as early planning documents insisted it should be, "freed from the impediments to change, such as rigid building codes, traditional property rights and elected political officials."[69]

Over the next four years, Disney's Imagineers would transform the wet wilderness into a tightly controlled environment. Working around the clock under bright spotlights, workers excavated 8 million cubic yards of fill to make way for the Mouse. To get rid of water, they carved out 40 miles of canals, 18 miles of levees, and 13 water-control structures. They drained the property's large lake, removing the decaying organic sludge that turned the water brown. They dug a two-hundred-acre lagoon to make Walt's blue-water entrance to the Magic Kingdom. They used the fill to raise the theme park 12 feet above its surroundings.[70]

Walt Disney World opened on October 1, 1971, after spending a total of $400 million on construction.[71] Three decades later, the Walt Disney Company, now the number two media conglomerate in the world, had made Central Florida home to four theme parks, two water parks, and scores of other thematic lodgings and attractions. It also was drawing far more visitors to the Sunshine State than its boosters ever dreamed—at last count more than 40 million a year.[72]

Like the generations of pitchmen, promoters, and developers who came before him, Walt Disney, with considerable help from Florida's elected leaders, imagined a paradise and made it come true. His dream took shape as an oasis: his blue lagoon. The canals that snake throughout the theme parks. Gently bobbing water rides for children. Soaking thrill rides for adults. Glittering, manufactured lakes reflecting patriotic fireworks that burst over the parks at night. And everywhere: fountains. In all, 85 of them grace the four parks. The most famous is Epcot's Dancing Waters, whose music-timed sprays can reach heights of up to one hundred feet from 264 jets.

A few years ago, Dancing Waters dried up.[73] Disney officials had to shut it down to comply with water restrictions as Florida faces a water-supply crisis—one that would seem impossible in a state so recently submerged.

Like the myth of Ponce de León's Fountain of Youth, like that of Cape Coral as "Water Wonderland," Walt Disney's oasis does not really exist.

You might call it a mirage.

2

Conspicuous Consumption

THEIR MAGNIFICENT ARCHES still stretching across parts of Europe from Italy to Istanbul, the Roman Empire's crumbling aqueducts stand as reminders of one of humankind's most enduring feats of engineering: the capture and carrying of water.

In his treatise on the water supply of ancient Rome, Sextus Julius Frontinus, the proud *curator aquarum,* or water commissioner, wrote that for 441 years the empire got by on natural water supplies. "The needs of Romans were satisfied by the water that they used to draw either from the [River] Tiber, from wells or from springs," the memory of which, he wrote around A.D. 100, "is still considered holy and revered."[1]

Indeed, most Roman cities had plenty of drinking water, with individual houses supplied by cisterns or wells. So why did the Romans spend five hundred years building thirty-one aqueducts, some as long as one hundred kilometers, throughout the empire? It was not out of necessity. They did so to sustain, and also to symbolize, a lifestyle of luxury, one based upon water.[2]

Emperors ordered aqueducts built or expanded primarily to supply water to the imperial public baths that were so central to the Roman way of life. But ostentation was a factor, too. Along with the baths, a generous water supply meant a city could have public water shows, glorious fountains, lush watered terraces, and other hydroluxuries.[3]

Take the ancient city of Perge, in what is now Turkey. There, water shot from an intricate, two-story aqueduct terminal before cascading into a canal that flowed down the middle of the main street and across the entire length of Perge. Adding to the pleasant sound and sight of moving water, the canal cooled the Mediterranean city. "Extravagances such as this surely reflect more than just a pride in the water system," writes classics scholar Trevor Hodge. "They reflect the particular form that took: an ostentatious insistence on abundance, or, as we would now call it, conspicuous consumption."[4]

DOWN THE DRAIN

In the modern world, Americans win the chariot race as the most conspicuous consumers of water. In all, the United States draws up 408 billion gallons of water a day, with crop irrigation by far the largest use. California, Texas, and Florida, where 50 percent of the nation's future population growth is forecast to occur, are the thirstiest. These three states use a quarter of all freshwater in the United States.[5] Managers for the nation's water supply stay awake at night worrying that we use more than nature replenishes. Water managers in thirty-six states believe they will see water shortages in the next decade. And that's without drought.[6]

But most Americans do not spend a moment fretting about the source or quality of their water. They expect it will gush out when they turn on the tap. It almost always does. The abundance, low cost, and cleanliness of water in the United States all help make Americans the heaviest water users on the planet. The average American uses more water each day than anyone in any other developed or undeveloped country.[7]

Each of us uses about 90 gallons of drinking water a day at home; each household about 107,000 gallons of water a year. Almost all of that is treated to meet federal drinking-water standards. We use it to flush toilets, water lawns, and wash dishes, clothes, and cars. More than half of all home water use in the United States goes to greening lawns and gardens. About 14 percent is never used at all. It leaks out of our pipes.[8]

Still, Americans care enough about water that they have made significant strides to conserve it. People are often surprised to learn that

U.S. water consumption has been on the decline for thirty years. Thanks to public awareness, modest water-conservation measures, and progressive pricing models that charge consumers based on how much they use, the country has seen particularly dramatic decreases in consumption since the 1980s. Nationally, average per capita use in 2000 was lower than it had been since the 1950s. And total freshwater withdrawals in 2000 were less than in 1975 despite population growth.[9]

That is not, unfortunately, the case in Florida. In the Sunshine State, both per-person consumption and total water withdrawals are on the upswing.[10] Perhaps because they see so much of it, Floridians view their supply of water as endless.

That too, of course, is not the case.

THE WATER STATE

Lawmakers designated Florida the Sunshine State in 1970, though car license tags carried the nickname for years before.[11] The legislature wanted to bolster the peninsula's balmy image. "Humid subtropical"—the state's official climate category, whose primary characteristic is hot, humid summers—was not quite so catchy.

In his book *Florida Weather,* geographer Morton Winsberg tells a tale of another Sunshine State: Queensland, Australia. The two Sunshine States are similar in climate. Both have stretches along their east coasts known as the Gold Coast and coastal roads called Highway 1. Each is home to a Miami, and a Palm Beach too.[12] But in Florida's case, the Sunshine State moniker is a misnomer. In the drier states of Arizona, California, and New Mexico, the sun is visible from the ground longer than anywhere in Florida, where clouds frequently obscure it.[13]

Floridians are not so much defined by the sun. They are defined, instead, by the water. It surrounds the peninsula on three sides. It seeps into the skin from the heavy, humid air. Depictions of Florida—whether by the most talented painters or the cheesiest promoters—are usually waterfront: vistas of the state's beaches, its wide bays, its 10,000 miles of rivers and streams, its 7,800 lakes.

Like ancient Rome, Florida sits in a water-rich part of the world; it is blessed with an extraordinary supply of groundwater; and it is home to hundreds of springs—seven hundred, to be precise. As much water as you can see in Florida, there is even more of it you cannot. More than 1 quadrillion gallons pulse through deep cracks and channels in the state's limestone core.[14]

But, like the Romans, Floridians are not satisfied with their natural water wonders. They like to show off, to play in, their artificial ones. Fountains grace the wealthiest city centers, the entrances of the best gated communities, and the best resort hotels. Canals cut through subdivisions so middle-class residents can enjoy the waterfront lifestyle of the rich. Florida is second only to California in its number of swimming pools, with 681,340 installed in 2004 alone.[15] Residents and tourists not content with beaches or pools can find more than a dozen water parks, where they can plunge down steep water slides or surf machine-made waves to a thumping beat of classic rock.

Typhoon Lagoon may not compare to a Roman city's waterworks or even, in the words of Frontinus, those "useless" wonders of the ancient Greeks: "With so many indispensable structures carrying so many aqueducts you may compare the idle pyramids or the other useless, although famous, works of the Greeks!" Frontinus brags in *De Aquaeductu.*[16] But the point is this: like the ancient Romans, Floridians are such conspicuous consumers of water that they can no longer get by on their bountiful natural supply.

THE STORY UNDERGROUND

By the turn of the twenty-first century, Florida's population had passed 15 million. The aboveground consequences were clear to anyone navigating the traffic-choked highways or the condo-crowded coasts. But the more insidious impacts were invisible. They were happening underground, in the vast Floridan Aquifer. The Floridan is one of the most productive aquifers in the world. It underlies 100,000 square miles of the southeastern United States, carrying groundwater through southern Alabama, eastern and southern Georgia, southeastern Mississippi, the bottom third of South Carolina, and most of Florida.[17]

In 2005, the U.S. Geological Survey reported on the cumulative effects of enormous groundwater withdrawals from the Floridan between 1950 and 2000. Geologists Richard Marella and Marian Berndt found alarming groundwater-level declines and saltwater intrusion throughout the aquifer. In northeastern Florida and eastern Georgia near Savannah, water levels since 1950 dropped at an average one-third to one-half foot per year. Along the coastal panhandle of Florida, water levels plummeted one hundred feet since 1950, causing utilities to punch wells farther and farther inland. Groundwater withdrawals sucked the life from numerous lakes, wetlands, and springs as well.[18] Kissengen Springs, a once-popular tourist attraction in Central Florida that bubbled up thirty cubic feet of groundwater each second, was the first major spring in the Sunshine State to completely dry up due to groundwater overpumping.[19]

Along the coasts of Georgia and South Carolina, large groundwater withdrawals in the Savannah and Hilton Head areas resulted in saltwater intrusion, which in turn caused high chloride concentrations in the remaining groundwater.

The scientists also reported grimly on groundwater quality. They found that throughout densely populated areas in Florida and Georgia, urban runoff and septic tank discharges tainted groundwater with nitrates and organic compounds. In Orlando, 240 drainage wells sent untreated storm water and urban runoff straight into the Upper Floridan.[20]

Throughout the Floridan Aquifer, groundwater was contaminated by land-use practices, too. Sinkholes and streams serve as funnels down which contaminants pour directly into groundwater. Chemical fertilizers, large numbers of farm animals, and septic tanks all cause nitrate pollution. Marella and Berndt found nitrates, as well as herbicides and pesticides, in springs and wells throughout North Florida. Nitrate concentrations in Manatee Springs near Gainesville, for example, had increased from 0.4 milligrams per liter in 1946 to more than 1.5 milligrams per liter in the late 1990s.[21]

Nitrates are notoriously bad for springs; they can turn a pane-clear blue grotto into pea soup, fill it with weeds. Elevated nitrate concentrations in rivers and springs also cause eutrophication, resulting in algal blooms and oxygen depletion that can lead to fish kills.[22] Nitrates pose dangers to humans, too, particularly to infants. In babies younger than six months old, nitrates can prevent blood from delivering oxygen to different parts of the body, causing sometimes-fatal "blue baby disease," which colors the sick baby's mouth, hands, and feet an obvious blue.[23]

The U.S. EPA's drinking-water standard allows 10 milligrams of nitrate per liter. Some advocacy groups advise against using water with levels exceeding 1.0 milligram per liter to mix infant formula.

Pollutants' harm is relative, depending on whether you're an adult or a tiny baby, a fish or a Florida spring pool. Even though it is safe to drink, groundwater with ten milligrams of nitrate would devastate the springs, reports Bruce Ritchie at the *Tallahassee Democrat* in a story about how Florida's nitrate standards are not nearly strong enough to keep its springs healthy. The standard "needs to be changed," springs biologist Jim Stevenson told Ritchie. But that could affect things like development, fertilizer use, and how wastewater plants operate. "That would influence a lot of land uses," Stevenson observed.[24]

DEVASTATING DROUGHT

Just as most Americans hardly think about water when they turn on the tap, the average Floridian did not really have to think about the state's groundwater problems, despite their enormity. But that changed in the year 2000. For many, the year was an unbearably long one. It was the driest in Florida's recorded history, the worst in a four-year drought that stretched from 1998 to 2001.[25] Nearly 1 million acres of wildfires burned throughout the state—despite a ban on outdoor fires that extended even to Boy Scout campfires.[26] Yawning sinkholes opened in yards, in the middle of highways. Thousands of private wells went dry. Lake beds turned to sand, sprouting ten-foot-tall weeds. In a Tampa warehouse, the Salvation Army stockpiled hundreds of cases of bottled water for families whose wells dried up. Florida's governor, the two-term Republican Jeb Bush, convened a "drought summit" of state leaders who drafted an emergency plan that called for trucking water to parched towns and cities, rationing supply to one gallon per person per day, and leasing portable desalination trucks similar to those used by the military during the Persian Gulf War. The governor called the drought "a crisis, not a potential crisis."[27]

Across the state, residents endured water restrictions. They could water their lawns or wash cars only once or twice a week. Power washing was banned in some communities; fountains ordered shut off in others. The shortage turned people surly, with neighbors calling special hotlines to tattle on wasteful neighbors and "water cops" patrolling subdivisions to look for violators. In South Florida, the water shortages were so severe that children were restricted in their use of "water-based recreation toys"[28]—like the ubiquitous yellow Slip N' Slide.

Lack of rainfall and groundwater pumping both can dry up lakes in Florida. At Lake Brooklyn in the Keystone Heights area of north-central Florida, docks stretch out into sand and grass.
(Photograph by Jon M. Fletcher, courtesy of *The Gainesville Sun.*)

Yet there was one group of Floridians under little pressure to curb water use: real estate developers. Florida has five water-management districts that oversee the state's water supply. The districts can permit or restrict water use, and they do. But the gubernatorial appointees who run the district boards are not inclined to quash economic development. While all five districts ordered water restrictions during the drought, overall water use actually climbed—in part because of new development. In the South Florida Water Management District, regulators say they were practically relieved when builders finally paved over the last possible piece of developable land in western Broward County. Finally, they could stop doling out large water permits. But then the builders surprised them by turning back to the east, and knocking down single-family neighborhoods to construct high-rise condominiums with even greater water needs.

In the sinkhole-prone Southwest Florida Water Management District, then-executive-director Sonny Vergara tried to explain to the *Lakeland Ledger* newspaper in parched Polk County how he could dole out water permits for new development with one hand and stop people from watering their lawns with the other. The drought, he said, was a short-term issue, while development was a long-term one. The biggest complaint he heard from the public, he said, "is that they're fined by governments for watering too much while the same governments are issuing permits for more building."[29]

Florida drinks up a total of 8.2 billion gallons of freshwater a day. The largest slice of the state's water pie—3.92 billion gallons, or 48 percent—goes to agricultural crops. Florida's farmers make the biggest withdrawals for irrigation east of the Mississippi River. In the year 2000, inefficient flood irrigation still quenched 45 percent of the state's farmland.[30] For most of Florida's modern history, the state's farmers were so revered that few dared question their water use. That trend has changed in recent years, though, as they have lost literal and political ground to home builders.

In almost every other part of the country where agriculture is being converted to housing, water consumption is going down. But just the opposite is happening in Florida. Even as total water use and per-person use dropped in the United States as a whole over the past few decades, both increased in Florida. The state's total freshwater withdrawals increased 46 percent between 1970 and 2000.[31] Per-person use climbed, too. In 1955, it was a little less than 140 gallons a day. Now, it is 174 gallons a day.[32] The upward trend has continued in recent, water-conscious years. Nationally, per-person water use dropped between 1995 and 2000, the latest period for which data are available. But in Florida, it increased—by 5 gallons per person per day.[33]

State water officials blame the four-year drought, saying it kept residents running their sprinklers in an effort to turn brown lawns green. Moreover, they maintain that it is unfair to compare Florida to the rest of the United States because of the Sunshine State's year-round climate. Up to 75 percent of domestic water use in Florida is for outdoor purposes, depending on the time of year.[34]

Indeed, Florida's green-grass culture does not look kindly on conservation. Just ask Sol Koppel, a retired computer programmer from New York. Koppel never had a yard until he was in his sixties. That is when he and his wife moved from Brooklyn to a community called Oakwood Village in the Tampa Bay area. In Brooklyn, he had never had more than a houseplant. Now, he had a huge lawn that he had to mow, weed, and water—even during times of severe water shortages.[35]

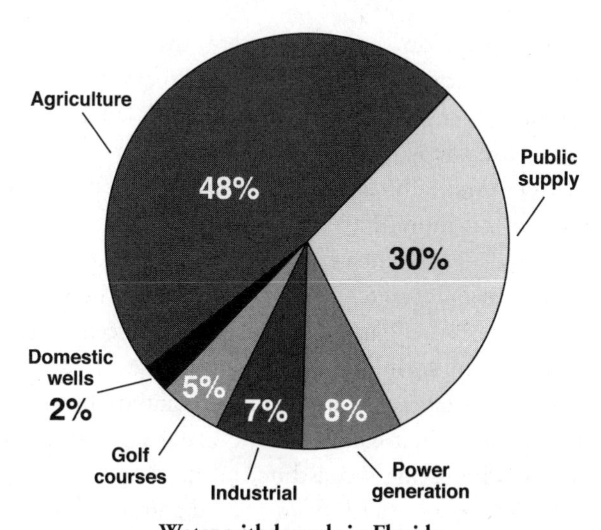

Water withdrawals in Florida.
(Data from 2000 U.S. Geological Survey. © Christopher Sheek.)

Koppel went to his local library to read up on drought-friendly landscaping, known as "xeriscaping." Then, piece by piece, he began to tear out his grass and plant drought-tolerant flowers, trees, and shrubs: Indian hawthorn, verbenas, and the like. Koppel's work earned him a native-gardening award from the University of Florida's Cooperative Extension Service. But it earned him the scorn of the Oakwood Village Homeowners Association. The association cited him for "noncompliance with deed restrictions" for not having grass. After a month-long fight, the mild-mannered Koppel reached a compromise with the Oakwood Village officers that let him keep some of his water-wise yard. He had to replace a 25-foot-by-8-foot swath with grass. "It was my first crime," he says. "Can you imagine? Not having enough grass."

Xeriscaping has become increasingly popular in Florida in recent years, and it is required in a few particularly water-stressed communities. But somehow, the greenness of grass has become a status symbol, so much so that some homeowners associations measure the shade of green—and ticket residents whose lawns are not bright enough. Some of the more notorious associations ticket residents for having yellow or brown lawns even during periods of drought, forcing them to choose between violating water restrictions and facing the wrath of their neighbors.

This copious irrigation, combined with the fact that the state is among the wettest in the United States, with at least 50 inches of rain a year in most places, makes Florida's lawns and landscapes the most watered anywhere in the world.

Green matters especially to golfers. Florida is home to more than 1,200 golf courses, far more than any other state in the nation.[36] At last count, they covered an estimated 134,000 acres, 90 percent of which were irrigated. Golf course acreage increased 58 percent between 1985 and 2000, while golf course irrigation increased 126 percent, again because dry conditions kept the sprinklers whirling.[37]

Nowhere are the golf greens more plentiful than in southwest Florida, where the Rosens dug their canals off the Caloosahatchee River in the 1950s and 1960s. The region's coastline, home to cities such as Naples and Fort Myers, has become a sea of vast gated communities, extra-large-lettered road signs, sparkling new healthcare facilities, and late-model Cadillacs. Until recently, anyone driving just a few miles inland could see that agriculture still ruled the region, with farmers dominating both land use and water use. Tomatoes, bell peppers, watermelons, and other thirsty crops covered most of the southwest interior all the way to Lake Okeechobee. But just as the farms replaced the wetlands in the twentieth century, the suburbs are replacing the farms in the twenty-first. Water managers say that by 2020 farmers will no longer be the biggest water users in southwest Florida.

For the first time anywhere in the United States, that honor will go to golf courses.[38]

CRISIS IN A LAND OF PLENTY

From the humanitarian standpoint, it is ludicrous to fret over water for golf courses and water for lawns—or to call a shortage of water for such things a "crisis." In the fall of 2005, Hurricane Katrina's drowning of New Orleans proved that wetlands drainage, development patterns, and engineered waterworks could have deadly consequences to Americans. But no true water "crises" exist in this nation, if your measure is the number of people dying for lack of clean water.

Water-related disease is the leading cause of sickness and death in the world, killing between 2 million and 5 million people a year. The vast majority of those deaths are children under five.[39] This was the rallying cry for the late Paul Simon, the bow-tie-wearing U. S. senator from Illinois. Around the world, 9,500 children die each *day* because of lack of water or, more frequently, diseases caused by polluted water.

"If one 747 plane filled with 350 children were to crash, killing all those on board, we would be mesmerized by the television and radio reports, and the story would fill the front pages of our newspapers," Simon wrote in his book *Tapped Out.* "Yet at least sixteen times that many children die each day for water-related reasons, but they do it quietly, and their stories rarely reach our living room TV sets and seldom even appear in the back pages of our newspapers."[40]

Around the world, more than a billion people do not have access to safe drinking water, and their numbers are growing. The World Health Organization reports that 1.1 billion people globally do not have access to "improved water supply," and more than 2.4 billion—40 percent of all people—lack basic sanitation.[41] By midcentury, the planet's population will increase from today's 6 billion to nearly 9 billion. Almost the entire increase will occur in countries already suffering water shortages.[42]

"The failure to provide safe drinking water and adequate sanitation services to all people is perhaps the greatest development failure of the twentieth century," says Dr. Peter H. Gleick, president of the California-based Pacific Institute for Studies in Development, Environment, and Security and one of the world's leading water scholars.

In 2000, the United Nations declared in one of its Millennium Development Goals that the international community would half, by 2015, the proportion of people without access to clean water. Other global summits have made similar vows. Instead, the trend is headed in the opposite direction.[43] The number of people without access to freshwater is expected to grow to between 2.6 billion and 3.1 billion by 2025. By that time, too, the number of countries facing water crises will grow from 25 today to between 36 and 40—most of those in Africa and western Asia.[44]

Florida, in the early years of the twenty-first century, was all about globalization—Jeb Bush trotting on trade missions around the globe, the state's public universities setting up global centers for this and for that. What would happen if a land of water-plenty put just a fraction of its wealth and science toward one of water scarcity? A Tampa cardiologist named Kiran Patel posed just such a question in 2005, when he made the largest single cash donation in the history of Florida's universities—with matching grants it would amount to $62.5 million—for a Center for Global Solutions at the University of South Florida. One of its research areas is water supply in the third world. "How much less water can our children drink?" asks Patel.[45] His own children live in Tampa. He was referring to the children of India.

The global water crisis demands America's attention for reasons beyond humanitarian aid. The lack of clean water and sanitation knocks more than $550 billion a year off world economic growth, or 1 percent of global gross domestic product.[46] At the same time, dwindling water supplies threaten violence in several hot spots: water is at the heart of tensions between Israelis and Palestinians; between Turkey and Syria; and along the Nile, Niger, and Zambezi rivers in Africa, just to name a few spots.[47]

The "crisis" of water supply in the United States is clearly incomparable to that of water shortages in Africa, China, and India. But whether viewed through an international, national, or Florida-centric lens, the basic facts are the same: The population skyrockets by the day. The amount of water available to the population stays the same. Whether at the global, continental, national, or state level, the problem is not really supply. It is timing and distribution.

Although water is the most widely occurring substance on Earth, only 2.53 percent of the planet's water is fresh; the rest is salt water. Of that slice of freshwater, two-thirds is trapped in glaciers and permanent snow cover.[48] So all in all, less than 1 percent of the world's water is available for humans. And its distribution is skewed. North and Central America have 15 percent of the world's freshwater and 8 percent of its population. South America has 26 percent of the water and 6 percent of the population. Australia has 5 percent of the water and less than 1 percent of the people. Europe, Africa, and Asia all have higher percentages of population and lower percentages of water. Asia is the most dramatic example, with 60 percent of the world's population and only 36 percent of the available water.[49]

Within continents, the issue is the same. Parts of Chile, Peru, and other South American countries have extreme shortages in some areas and severe flooding in others. In North America, it is the same. Here in the United States, the Great Lakes region holds 95 percent of the country's fresh surface water but only 8 percent of its population. Within states, it is the same. In Florida, where the primary source of freshwater is groundwater recharged by rainfall, most of the rain falls in the north of the state, while most of the people live in the south.

This is one weird fact about Florida's weather you may not have learned on the Weather Channel. The state is split in half, just above Orlando, by a horizontal line that geologists call the "hydrologic divide." Only 44 percent of Florida's rain falls south of the line, yet 78 percent of the state's permanent population has settled south of it. In her book *Florida Waters,* Florida State University professor Elizabeth D. Purdum explains how neither surface water nor groundwater crosses this line, making South Florida an island wholly dependent upon rainfall.[50]

At the beginning of the nineteenth century, more than half of Florida's 36 million acre peninsula was submerged. By the beginning of the twenty-first, vast drainage and reckless groundwater pumping finally proved the limits of the state's water bounty. Floridians were getting rid of more groundwater every day than the state's water cycle could replace.[51] In a region lashed by hurricanes each summer, in a state that gets 150 billion gallons of rainfall each day, and another 26 billion gallons a day from rivers that flow from Georgia and Alabama,[52] it was an astonishing feat.

3

Drained & Diverted

If the aqueducts are windows into water in the ancient world, the best place to contemplate water in the United States today may be from a sidewalk that runs along U.S. Highway 93 in eastern Nevada as it crosses over the Hoover Dam. To fathom how water could be disappearing in the verdant East, it's worth seeing where it went in the arid West. To figure out increasing conflicts over water in the East, it helps to look at who won the battles in the West. Ultimately, it doesn't matter whether you're talking about ancient Rome, the American West, or modern Florida. Those who control the water control the destiny of a place and its people.

Along that Nevada sidewalk, from a half bowl at the top of Hoover Dam, an arc of concrete narrows 726.4 feet to a dull point at the bottom, inspiring vertigo as much as awe. Built in the 1930s, during the Great Depression, the dam was known for its engineering superlatives: highest dam ever, costliest water project ever, largest power plant ever.[1]

Atop the structure, art deco motifs, including two enormous statues of winged men, are redolent of the Great Society. In a U-shaped building at the bottom of the dam, seventeen mammoth hydraulic turbines spin in cavernous rooms furnished with Industrial Age gauges and switches. No computers can be seen. Visitors feel the thrum of the turbines and smell the lubricating oil. In the information age, the place feels like a relic. But it still provides flood control for California's Imperial Valley, irrigates more than a million acres of agricultural land, supplies drinking water to more than 18 million people, and generates about 4 billion kilowatt-hours of energy—enough for 500,000 homes—every year.

By harnessing the Colorado and virtually every other river in the region, the western United States supports 40 million acres of agricultural crops, the most productive in the country. Taming the rivers also brought life to some of the most vibrant and populous cities in America. These crops and cities sprout where they arguably should not exist—in a region as dry as North Africa.[2]

In all, Americans have built some 76,000 dams, among them more than 2,000 hydroelectric power dams like the Hoover and the Glen Canyon dams on the upper Colorado in Arizona.[3] Without the Hoover, without the Glen Canyon Dam and its 27 million acre-feet Lake Powell, without the tallest—at 770 feet—Oroville Dam on the Feather River in northern California, there would be no West as we know it. No Los Angeles. No San Francisco. No Phoenix. No Las Vegas. The Hoover also inspired a global binge of dam building that continues today. In 1900, no dam in the world reached higher than 15 meters. By 1950 there were 5,270 of them; two in China. Thirty years later there were 36,562 of them; 18,820 in China.[4]

In 1893, the historian Frederick Jackson Turner presented his mythic frontier theory of the United States. Americans' continual battle with the primitive conditions of the western frontier, like aridity, he argued, gave the nation a one-of-a-kind culture of individualism, self-reliance, and diffused power. "This perennial rebirth, this fluidity of American life, this expansion westward with its new opportunities, its continuous touch with the simplicity of primitive society, furnishes the forces dominating American character," Jackson said in his celebrated essay "The Significance of the Frontier in American History." "The true point of view in the history of this nation is not the Atlantic coast, it is the Great West."[5]

A century later, Donald Worster, an eminent environmental historian of the West, turned Jackson's frontier theory on its head. Worster pronounced the West a hydraulic empire, one that led not to diffused power but to ultimate power concentrated in the hands of the few who control the region's vast waterworks.[6] Men like U.S. Representative Wayne Aspinall of Colorado spent an entire congressional career—his lasted from 1949 to 1973—making sure they brought water projects home to their districts.

Few would dispute the business-development and social benefits of dams, from hydropower to flood control. In January 1997, the city of Reno was saved from "extreme disaster," according to scientists, because managed release of water from upstream reservoirs, made possible by dams, prevented a one-thousand-year flood. Still the conventional wisdom on dams in this country has slowly changed since 1978, when President Jimmy Carter vetoed the entire U.S. appropriations bill to protest what had become increasingly wasteful, pork-barrel dam projects that cost millions and benefit few. (Congress overrode the veto.) Of the projects on Carter's hit list, one would return 5 cents in economic benefits for every taxpayer dollar invested. One offered irrigation farmers subsidies worth more than $1 million each. Another, a huge California dam, would cost more than the Hoover, Shasta, Glen Canyon, Bonneville, and Grand Coulee projects combined.[7]

The heavily subsidized projects also have devastating ecological consequences. Dams drown wildlife habitat under reservoirs and block annual migrations of salmon and other fish. In California alone, 80 percent of the salmon and steelhead populations have been wiped out since the 1950s. And if you care about nature's majesty, dams submerge it. "You once had to backpack or paddle a raft to see the slickrock, sandstone arches and 1,000-foot cliffs of Glen Canyon," *Denver Post* environmental reporter Mike Obmascik says of man-made Lake Powell. "Now you can see it drunk and chain-smoking on a lounge chair from the stern of a Boston Whaler."[8]

In the twenty-first century, America's love affair with dams seems to be coming to an end. In 2005, 56 dams in eleven states were scheduled for demolition in the name of river restoration. But the story of the Animas–La Plata dam now under construction in southwestern Colorado shows that when it comes to waterworks, politics can overpower science, economics, and even common sense. It is a tale worth recounting as eastern states muscle in for bigger and bigger federal water-supply projects.

Animas–La Plata is a legacy of Aspinall, who ruled the House Interior Committee for more than a dozen years. In 1978, Carter's vetoes included the (now) $710 million Animas–La Plata, which had been authorized a decade before. The dam and irrigation project would require huge taxpayer subsidies, with almost three-quarters of its cost going to provide irrigation water to farmers who would repay only 3 percent of the construction expense. The Department of Interior's inspector general called it economically unfeasible. Other government auditors said it would deliver less than 40 cents in benefits for every dollar spent.[9]

The *Washington Post* dug up an ad from the *Durango Herald* of Colorado that captured in just five sentences the problem with all federal taxpayers footing the bill for pet water-supply projects. "WHY WE SHOULD SUPPORT THE ANIMAS-LA PLATA PROJECT," the ad began. "BECAUSE SOMEONE ELSE IS PAYING MOST OF THE TAB! We get the water. We get the reservoir. They pay the bill."[10]

After a half century of debate, in August 2005, the U.S. Bureau of Reclamation broke ground on Animas–La Plata. The same government that has torn down more than four hundred dams nationwide in recent years because of environmental, safety, and other concerns is building a brand-new one that makes no economic sense. Animas–La Plata is supposed to be finished in 2008; its 120,000-acre-foot reservoir filled by 2011.[11]

It ultimately came down to politics. Marc Reisner, a gifted environmental writer who died of cancer in 2000 at the age of fifty-one, famously put it this way: "Congress without water projects would be like an engine without oil. It would simply seize up."[12]

In *Cadillac Desert,* his classic water history of the West published in 1986, Reisner chronicled legendary battles between the U.S. Bureau of Reclamation and the U.S. Army Corps of Engineers over which federal agency would build and control western dams. The bureau ultimately won the war, and the spoils. The bureau was created in 1902 under President Theodore Roosevelt as part of the Newlands Act. Named for its chief sponsor, U.S. Representative Francis G. Newlands of Nevada, the act created a fund from the sale of public lands in sixteen western states to develop irrigation projects. Settlers benefiting from the projects were supposed to repay their costs, creating a permanent revolving fund. The pay-back part never worked as it should. But the Newlands Act ensured that the federal government would control the large-scale irrigation projects of the West. By 2005, the Bureau of Reclamation had built so many dams, reservoirs, and irrigation ditches that it had become the largest wholesale water supplier in the United States.[13]

While the bureau made itself essential supplying water for people, the Army Corps carved a niche in keeping people safe from water. Corps history dates back to the Revolutionary War. Congress authorized the Continental army's first chief engineer to build fortifications around Boston at Bunker Hill. The agency's most famous project is perhaps the Tennessee Valley Authority (TVA), a hallmark of President Franklin Roosevelt's public works agenda that tamed the wild Tennessee River and its tributaries. The TVA's forty-seven dams, most built after a bruising battle over what Roosevelt's foes called his "socialism," generate power for a once-underdeveloped region, help stem flooding, and fill reservoirs for water sports. They ease navigation along the river, sparking commerce and manufacturing.[14]

The TVA also is $30 billion in debt, has advocated ruinous activities such as coal strip-mining in the Appalachia region, and is the nation's worst violator of the Clean Air Act. And it is so politically powerful that when it wanted to build a dam called Tellico that a cabinet-level committee unanimously opposed on economic grounds and the U.S. Supreme Court stopped under the Endangered Species Act, Congress voted to exempt the dam from the act and other laws. The dam got built.[15]

THE ARMY CORPS AND THE EVERGLADES

During the deadly Florida hurricane of 1928, twelve feet of water had topped Lake Okeechobee's dikes and killed more than 2,000 people. Congress called down the Army Corps to build the Herbert Hoover Dike, a 140-mile earthen dam that surrounds Lake Okeechobee. In the 1930s, drought caused a different sort of disaster. South Florida's numerous canals were having their intended effect of drying up the Everglades muck, composed of peat soils that took thousands of years to form. During the drought, dried-up areas of peat would catch fire and burn for months, creating a dense pall of smoke over surrounding towns. On the east coast, the lowered water levels were causing a new crisis. Residents' wells began to draw salty water, a problem they fixed by moving the wells farther and farther inland.[16] That trend continues to this day, although now it takes place up and down the entire peninsula, from the tip of South Florida all the way to the once-desolate, now fast-developing panhandle.

In 1947 and 1948, disastrous floods returned, making it increasingly clear that draining the entire Everglades would lead to disaster for those who ultimately settled there. Americans were finally convinced of the folly of draining the great marsh. The Hoover Dam, TVA, and other grand waterworks of the hydraulic empire showed them it would be easy enough to simply tame it.

Congress's Central and South Florida Project ordered the Corps to replumb the entire bottom half of the state to provide flood protection and freshwater for urban and agricultural lands. The engineers chopped up the southward-flowing Everglades with 1,000 miles of canals and 720 miles of levees, controlling the flow with sixteen pump stations and two hundred gates and other concrete-and-steel structures. They built towering gates at Lake Okeechobee so the state's water managers, then the Central and South Florida Flood Control District, could force water into the sea when floods threatened. They erected barriers between the sea and canals to block intruding saltwater. Finally, they turned a meandering, 90-mile-long river called the Kissimmee into a 52-mile canal. The ramrod-straight Canal 38, called C-38 for short, drained the floodplain marshes of the river and its headwater lakes to the north near Orlando.[17]

At the time, most Americans marveled at such triumph of man over nature. But fifty years later, half the Everglades were gone, and people realized what else they lost. Whether you care a whit for the panthers or the plume birds, the Everglades' most critical job was storing freshwater for a land surrounded by the sea.

Now that the nation's environmental ethic has shifted, it is easy to blame the Corps for the damage done in Florida and other places where its engineers dynamited and dredged. It is also easy to forget that the Corps' waterworks in the Everglades continue to save lives. Every hurricane season, Hoover Dike protects southeast Floridians from disastrous, and deadly, flooding. And it will, as long as it continues to hold. In May 2006, a panel of civil-engineering experts warned the dike is prone to collapse, presenting "a grave and imminent danger to the people and the environment of South Florida." A breach could imperil tens of thousands of people, inflict tens of billions of dollars in damages, and contaminate the region's drinking water. Without immediate intervention, the report said, the dike had a 1-in-6 chance of failing in any given year.[18]

Army Corps officials bristled at the state-commissioned report, and said it underscored their engineers' work to restore the Everglades while

Florida

(© Christopher Sheek)

providing flood control to the people of South Florida. Indeed, today's Corps is an agency at cross-purposes, trying to preserve its proud engineering past while forging a new role in environmental restoration. The results can be schizophrenic. Corps dams control flooding, but the agency's wetlands-drainage programs cause it. From Florida to New York, the Corps replaces beach sand washed away because of development it approved. The agency's attempts to control flooding have made flooding worse by inducing people to build in high-risk areas.[19]

In the Chesapeake Bay, the Corps is restoring islands, wetlands, and oyster bars at the same time its dredging activities are harming them.[20] In the Great Lakes, one arm of the agency wants to deepen and widen the

St. Lawrence River channel for larger and more ocean-going ships, while others in the same agency battle the exotic species that enter the Great Lakes on such ships.[21]

From the Everglades to New Orleans, the Corps is spending billions of taxpayer dollars to fix its own failures. The agency is directing a $4 billion plan to restore the Missouri River that it dammed and channeled to ease navigation, generate hydropower, and reduce flooding. It is spending $5.7 billion to restore fish and wildlife habitat along the upper Mississippi River destroyed by its dams, levees, and other engineering marvels that hold back water.

In South Florida, the Corps and the South Florida Water Management District are spending $10.5 billion to fix their own drainage and diversion projects that have destroyed the Everglades. At the same time, both agencies are permitting massive new developments that encroach on the ecosystem from all sides.[22] In 2005, for example, Corps regulators suspended a permit they had granted to a company called Atlantic Civil to fill 1,000 acres of sensitive wetlands in a rural area southwest of Miami, near crucial elements of the Everglades restoration plan. Atlantic Civil had applied to fill the wetlands for "agricultural purposes." But the plans it submitted to state agencies revealed these sorts of crops: 6,000 housing units, 300,000 square feet of retail space, 90,000 square feet of office space, three schools, one 1,800-seat theater, and so on. The regulators stopped the permit and explained that they feared its impact on Everglades restoration, that it needed "more detailed examination," and that it required public notice. Twenty days later, they inexplicably reinstated it.[23]

Diverted in the West. Drained in the East. In each region of the United States, Americans have altered the natural flow of their water, often with unintended results. Dredging and levees along the Mississippi River, the largest waterway in North America, have hastened the crumbling of more than 1,000 square miles of Louisiana into the Gulf of Mexico. Agricultural and industrial pollution flowing down the river's 2,350 miles have led to a growing, annual "dead zone" in the Gulf; an area of 7,000 square miles bereft of oxygen, and therefore life. New York City's water system, once one of the highest quality in the nation, is on the decline because of agricultural and urban-growth pressures in the Croton and Catskill/Delaware watersheds. From there, water barrels down two huge tunnels that supply 9 million New Yorkers with 1.5 billion gallons every day.

The richest source of water in the United States, the Great Lakes, is dogged by pollution, the spread of nonnative species that raid the five lakes' food chain, and fast-disappearing wetlands and habitat on most of their shorelines. More than two-thirds of the lakes' natural wetlands have been filled in or drained for development.

The Great Lakes hold a fifth of all the fresh surface water on the planet. Together with the St. Lawrence River, they form a navigational system as big as the Atlantic Ocean is wide. Millions of people depend on them for drinking water, manufacturing, shipping, recreation, fishing, tourism, and irrigation. While cleaned up considerably since the Industrial Age, the lakes in recent decades have become vulnerable to a new threat. Outsiders keep popping up with schemes to pipe, ship, or pump water out of the basin. So far, plans to divert Great Lakes water to keep Mississippi barge traffic afloat, to irrigate the Great Plains, and to ease water shortages in Asia have all been stymied. But midwestern leaders think it is inevitable that someday they will be targeted by water-needy states—such as California, Texas, or Florida—that have more political clout.[24]

A population boom in America's Sunbelt states has turned the country's power base upside-down. In 1940, northeastern and midwestern states had 251 seats in the House, compared with 184 for states in the South and West. Today, southern and western states have a 252–183 edge, one that in all likelihood will continue to widen. Southern and western states are growing so much faster than the rest of the nation that several will grab House seats from the Northeast and Midwest when Congress is reapportioned in 2010. Demographers project that Florida and Texas could each gain as many as three House seats. Ohio and New York could each lose as many as two.[25]

Many observers dismiss the idea that Great Lakes water will someday be piped south, or shipped to other points on the globe, to ease water shortages elsewhere. The late Senator Simon of Illinois, who knew water issues better than almost anyone else in Washington, was not one of them. "Inevitably, if we don't find answers, we are going to have to be piping water from the Great Lakes to the southwest portion of the United States," Simon said in 2000. "And when people of Illinois say that they would never tolerate that, well, senators from Illinois who want to get projects approved had better go along with high priorities for senators from the southwest states or they won't get projects approved for Illinois. It's not that complicated when it gets to the United States Senate."[26]

For better or for worse, the congressional power shift means more big water projects throughout the southern half of the United States. The multibillion-dollar projects are what Reisner famously called "the grease gun that lubricates the nation's legislative machinery."[27]

As the Hydraulic Age of the twentieth century gave way to the Restoration Age of the twenty-first, the largest water project in the nation (and in the world) aimed to "fix" the Florida Everglades. It was something Americans had been trying to do, in one way or another, for more than one hundred years.

"THE DESTRUCTION OF FLORIDA IS A PRICE TOO HIGH"

The draining and paving over of Florida's wetlands in the first half of the twentieth century was carried out unwittingly, to the extent that most Americans believed the only good swamp was a drained swamp. The rampant draining and filling in the second half of the century was the crime (literally, in some cases) because Floridians, by then, knew what they were doing. They passed aggressive laws to stop the wholesale destruction of natural Florida. Despite their best intentions, and no matter which political party was in power, it continued almost unabated.

From 1950 to 1970, Florida's population more than doubled, from 2.8 million to 6.8 million.[28] During 1970 and 1971, Floridians suffered through the most severe drought the state had seen since the grim 1930s. Water shortages led to rationing. South Florida's cities, farms, and Everglades National Park were fighting fiercely over Lake Okeechobee's dwindling water supply. Five hundred thousand acres of the Everglades burned. Day after day, week after week, the southern third of the state was covered with smoke so heavy that "everyone was constantly aware of it."[29]

It was clear that Florida's developers and farmers, along with the government's engineers, had gotten rid of too much water. As housing developments and farms rapidly took the place of wetlands that once stored water, the natural system no longer could hold enough to supply those houses and farms, especially during droughts. Population pressure exacerbated water problems, particularly saltwater intrusion and pollution from sewage and industrial and agricultural wastes. Just as it had in the 1930s, drought led Floridians to rethink the pace and manner of development and the ruin of the wetlands and the other natural resources.

Perhaps to counter the boisterous Claude Roy Kirk Jr., Floridians in the fall of 1970 elected the church-pew-straight Reubin O'Donovan Askew (1971–79), a moderate Democrat state senator from Pensacola, as Florida's thirty-seventh governor. In his inaugural address on January 5, 1971, Askew made clear the importance the environment would take in his administration. The first topic he raised was corporate tax reform, getting rid of "special tax favors to the politically influential" to raise money to pay for the impacts of rapid growth. The Florida legislature that very year passed a corporate income tax. (Over time, state lawmakers created so many exemptions, deductions, and other loopholes that by 2003 98 percent of Florida's 1.5 million companies, including huge corporations such as Carnival and Verizon, were not paying a cent of the tax. The loopholes now cost Florida about a billion dollars in tax revenue each year.[30])

Second, Askew said he wanted to reverse the trend of ecological destruction in Florida. He tapped a new environmental argument, one that would grow increasingly effective in the state, as well as other parts of the country in the business of selling vacations. Florida's economic health, he said, was directly linked to the health of its environment. "We must ensure our continuing economic prosperity, but the price we pay for growth must be carefully evaluated," he said. "We must recognize that the destruction of Florida is a price too high. *Ecological destruction in Florida is nothing less than economic suicide.*"[31]

Askew was soon on the cover of *Time* magazine, along with Governors John West of South Carolina, A. Linwood Holton of Virginia, Dale Bumpers of Arkansas, and Jimmy Carter of Georgia, in a story on "the new breed of Southern governors." *Time* declared that the five had "departed from the old practices of racism and provincialism and favor economic development balanced with environmental sensitivity."[32]

Indeed, Askew led Florida, in the 1970s, to pass some of the most aggressive land-use and water laws in the nation. The Environmental Land and Water Management Act gave the state authority to set development standards and guidelines for major projects like large housing developments and shopping centers. The Land Conservation Act authorized $240 million worth of bonds to buy up environmentally endangered and recreation lands. The Water Resources Act created a comprehensive system to conserve and protect Florida's water while maximizing its use. It carved the state into "water-management districts" drawn along hydrologic basins. The districts would tax land, regulate drainage, and dole out water permits for large uses.

The Water Resources Act declared Florida's water "a public resource benefiting the entire state" that should be "managed on a state and regional basis . . . so as to meet all reasonable-beneficial uses."[33] This made Florida's water law unique in the nation—protecting the public interest to a greater degree than most states. In western states, generally, water is governed by a doctrine called "prior appropriation": whoever was using the water first has first right to it. In the eastern states, water more often falls under riparian rights. That is, landowners have the right to the water their land touches. In Florida, the five water districts grant permits to users (at least in theory) based on evidence of three things: that the use will not harm other existing users, that it is "reasonable and beneficial," and that it is consistent with the public interest. The districts also can require compliance with additional environmental standards. Contrast that with the "Rule of Capture" in Texas, where people can take as much water as they want from under their own land, even if it impacts neighbors.

The suite of environmental laws was a sharp break with the past. Florida's new Comprehensive Planning Act, for example, "was a surprising change, because prior to that time, planning was like socialism right out of Moscow," remembered the House Speaker, Richard Pettigrew.[34] But the problem in Florida was not and is not the laws on the books. It's the way local elected officials make exceptions to them, the way savvy land-use lawyers and others get around them. Ultimately, newcomers were moving to Florida so fast that the progressive laws barely dented environmental devastation. Especially to the wetlands that clean and store Florida's water. From 1970 to 1980, the population jumped from 6.8 million to 9.7 million.[35] During that time, Florida had among the most extensive wetlands loss in the nation.[36] The U.S. Fish and Wildlife Service found that in the eight years between 1972 and 1980, South Florida drained 23,767 acres of wetlands for agriculture and 655 acres for urban development. During the same period, 24,539 farm acres in South Florida were converted to subdivisions.[37]

That pattern continues today. To see it, you can pay $3.50 and take a lonely elevator ride up twenty-two stories of concrete and steel in the Florida Citrus Tower north of Orlando. Erected in 1956 to offer tourists a panoramic view of 17 million orange and grapefruit trees, the tower today is little more than a memorial to the lush groves that inspired it. The view from the observation platform includes but one small patch of oranges. It is overgrown, and blocked by a large "for sale" sign. Around it, in every direction, housing subdivisions with names like Orange Tree stretch to the horizon. Deadly freezes seem to chase Florida's citrus growers, already squeezed by urbanization and globalization. Over the past twenty years, the industry has relocated farther and farther south onto drained land in the counties surrounding Lake Okeechobee, where acreage is cheaper and freezes less severe.[38]

While farmers use far more water than the general public, growth and development drive the fate of Florida's groundwater. Water use for farms has remained static over decades, while public use has grown exponentially and will eventually outpace agriculture. Water withdrawals for public supply in Florida increased 1,330 percent between 1950 and 2000. Over the same period, the population grew 475 percent, from 2.77 million to 16 million.[39] Some agricultural lands, such as cattle ranches, preserve wetlands and recharge areas; some are good candidates for future environmental restoration projects. But miles of paved surfaces like parking lots, highways, and driveways wreak havoc on the hydrological system by diverting the rain that recharges aquifers.[40] "Nature has an amazing resilience, an amazing capacity to rebound from the evils that man bestows upon it," says former U.S. senator Bob Smith, a New Hampshire Republican who has retired to Florida and leads a nonprofit called the Everglades Foundation. "But there is a point of no return, and that's development."[41]

In Florida, like many parts of America, economic development trumps environmental protection. This is true whether the Democrats or the Republicans are in power. When it came time to put Florida's new land and water laws to task, the state was suffering economically, along with the rest of the nation, in the recession of 1972–74. Despite considerable evidence to the contrary, the development community complained about "no growth" and "no jobs," in the words of Pensacola Realtor Theo D. Baars Jr.[42] Environmentalists charged the ambitious laws were dampened in the rule-writing and implementation process. Governor Askew admitted you "had to face the reality of job-creation."[43]

Then and now, the only surefire way to protect land and water is to buy them outright. Askew's land-conservation law was the saving grace of the 1970s. The state had begun acquiring land in the mid-1920s and had a formal program since the 1960s, funded by a 5 percent tax on recreational items dubbed "the bathing suit tax." Askew's $240 million bond issue, overwhelmingly approved by voters, further built a land-acquisition program that would become the largest in the world with $4 billion to spend—now called Florida Forever. Led by Askew and a young U.S. senator named Lawton Chiles, Florida and the federal government used the program to leverage a crucial wetlands buy: what is now the 729,000-acre Big Cypress National Preserve. A magical cypress stand to the west called the Fakahatchee had been picked up cheap by the Rosen brothers and subdivided into lots. Thanks to land preservation it is home, not to thousands of ranchettes and their residents, but to huge cypress trees, to tiny, rare orchids, and to endangered creatures like the wood stork and the panther.

The rest of South Florida was "going down the tube." At least that is what the *Sports Illustrated* swimsuit issue (cover: Christie Brinkley) declared in 1981, in a story that said, "In no state is the environment being wrecked faster and on a larger scale." The article reported how South Florida's mullet catch had plummeted 90 percent in five years, how the old flood-control district canals were swarming with coliform bacteria, how the Chain of Lakes above Okeechobee was collapsing.[44]

Other parts of the state were not faring much better. Decades of pollution and dredging for waterfront development around Tampa Bay, for example, had wiped out the sea-grass meadows and the natural shoreline. The bay's once-bountiful fishery was all but gone. Even the stately homes lining Bayshore Boulevard were not immune from environmental blight—especially on hot summer days. Residents claimed the stench from rotting algae fueled by the barely treated sewage that poured unchecked into Tampa Bay was tarnishing the silver.[45]

In 1982, the Congressional Office of Technology Assessment asked a University of Florida researcher named Richard Hamann to report on the consequences of a dried-up South Florida. His was a prescient warning—especially reading it twenty-five years later, when Florida's developers were pressuring politicians to find them more water. Much of the drainage and diversion carried out in the name of economic development in the twentieth century was foiling economic development in the twenty-first. Disappearing water supply and pollution were of equal concern. Water pumped from Lake Okeechobee was fouling estuaries in the coastal counties to the east and west; business leaders were as distraught as environmentalists: "We now recognize that our very existence is being threatened," Steve Greenstein of the Sanibel Chamber of Commerce complained in the fall of 2005 of the lake's releases into the Caloosahatchee River.[46]

Nearly a quarter century before, Hamann tried to sound the alarm, detailing eight of the most worrisome consequences of draining South Florida. He threw in a disturbing ninth that was still just a theory. The first was loss of water storage. Drainage for development, along with the replumbing of the Everglades, meant there was no place for the region's rain to go—except through canals and out to sea. "The loss of water storage capacity is of particular concern in view of the rising demand for consumptive use of water," Hamann wrote. He estimated South Florida could be home to as many as 7 million people by 2010, and that water demand there could reach as high as 2 billion gallons a day. The region was home to 7 million people by 2005; its daily water demand more than 5 billion gallons.

The second consequence was loss of organic soils, which also store water, just like a sponge. One inch of rainfall can raise the water table half a foot in the peat. Hamann also warned of severe saltwater intrusion, an increase in both the incidence and intensity of wildfires, loss of flood control, declining water quality, loss of fish and wildlife habitat, and reduced production in the estuaries. All would become headlines in the decades that followed.

And number nine? On this, Hamann was less certain. But scientists were beginning to suspect that wiping out so many wetlands could impact local *climate.* Large water bodies, they knew, helped prevent damaging frost. Could draining the Everglades make freezes more severe? What about rainfall? Could wiping out wetlands reduce summer thunderstorms by cutting the evapotranspiration that triggers the release of moisture from clouds as they move inland from the sea?[47]

Those theories, it turns out, were right as rain. By draining their state, Floridians were not just altering the supply and quality of their water. They were changing the weather.

4

The Wetlands & the Weather

ILLOGICALLY, THE HEART of climatology research in the United States, including hurricane research, is at Colorado State University in land-locked Fort Collins. There, a meteorology professor named Roger Pielke Sr. juggles his graduate students, his contributions to scientific journals, and his role as the state of Colorado's climatologist as he patiently tries to explain his views on global warming to journalists. But that story comes later.

Pielke began his career with the National Oceanic and Atmospheric Administration's Experimental Meteorology Lab in Miami, where in the early 1970s he worked on a government cloud-seeding project. Pielke's job was to develop a model of Florida's weather that NOAA could use to figure out if its cloud-seeding experiments were increasing rainfall. (They were not.) While the feds' foray into weather modification failed, Pielke continued to build upon his model. Fifteen years ago, he consolidated it with another developed by colleagues at Colorado State. They called the combined tool the Regional Atmospheric Modeling System, or RAMS.

In 2004, Pielke and his colleagues published two journal articles based on RAMS that drew this remarkable conclusion: by changing the land in Florida, people were changing local climates. Massive drainage of wetlands in particular, they found, could lead to decreased rainfall in the summers and more-severe freezes in the winters.[1] As they dried up the land, Floridians dried up the rain.

Deb Willard is a paleobotanist with the U.S. Geological Survey, but you might think of her as a nature detective. She created pre-1900 land-cover maps of South Florida using centuries-old clues such as pollen trapped in layers of soil. The scientists fed those maps, along with modern-day land-cover data from NASA's Landsat 5 satellite, into RAMS. That let them generate weather patterns for particular dates using either current or historic land cover.[2]

To study summer rain and heat, the team ran simulations for three separate summer-time dates. In each case, it found that when averaged across South Florida, modeled rainfall was 10 percent to 12 percent less when current land conditions replaced the pre-1900s natural vegetation. Temperatures were affected as well. Urbanization hiked maximum daily temperatures by several degrees, particularly around Miami and Fort Lauderdale. The thermometer climbed highest in the center of the peninsula, where daytime highs were up to four degrees Celsius warmer under the present-day land conditions. The scientists checked their simulations against actual summer-time rainfall data available from South Florida stations between 1924 and 2000. Though the station data were limited, those, too, revealed an average rainfall decrease of 12 percent over the seventy-five-year period.[3]

Pielke and his coauthors, Curtis H. Marshall of NOAA and Louis T. Steyaert of NASA, traced the changing rainfall patterns to drainage of the Kissimmee River basin floodplain. Inundated with water prior to 1900, the basin would have provided more moisture as fuel for Florida's summertime showers. Without the water, the showers were weakened over the state's interior.[4]

If the rain drying up were not bad news enough, Pielke and his colleagues next turned their attention to freezes. Using the pair of land-cover maps, the scientists honed in on a January 1997 freeze that had not been predicted and severely damaged South Florida's winter vegetable and citrus crops. They simulated the freeze with both data sets and found that under the pre-1900s natural land cover, "a persistent heat flux from wetlands was sufficient to hold the simulated temperature above freezing throughout the night."[5] Under current land cover, the simulation produced minimum temperatures that were both colder and below freezing for longer periods—particularly in areas to the south and southwest of Lake Okeechobee that had been drained for high-density cultivation of winter vegetables, sugarcane, and citrus.[6]

The study offers an answer to the long-troubling question of why freezes seem to chase Florida's citrus farmers. The state's citrus industry started in the nineteenth century in northeast Florida around Jacksonville. Deadly freezes kept farmers relocating farther and farther south, from their heyday along Central Florida's Indian River ridge in the mid-twentieth century to their current concentration in the counties around Lake Okeechobee. Ironically, says Pielke, when farmers drain wetlands for crops, they increase the likelihood that freezes will harm those very crops.

Florida's own climatologist is a white-bearded Santa look-alike named James O'Brien. He is an expert in the weather phenomenon El Niño, the name of a warm ocean current that appears irregularly off the north Peruvian coast and causes weather aberrations throughout the world, particularly in Florida. (El Niño is a good thing for Florida because its winds weaken hurricanes.) O'Brien wears bolo ties and loves bass fishing. He has a mildly annoying trait that gives him away as one of those most hardcore of bass fishermen. He will not reveal where he caught certain lunkers pictured on the walls of his office in a sprawling research park south of Tallahassee. "That one came from a large lake north of Orlando," he demurs. "That's as much as I can say."

There is one point on which O'Brien does not demur. Pielke, he says, is right that wetlands drainage is changing local weather. Scientists have long noted the worldwide "heat-island" phenomenon. Urban and suburban areas stay 2 to 10 degrees Fahrenheit hotter than nearby rural areas. The reasons: more asphalt, less water and shade, and other heat generators like car exhaust. A similar phenomenon affects rainfall. Crunching a century of precipitation data on his computers at Florida State University's Center for Ocean-Atmospheric Prediction Studies, O'Brien creates charts from cities throughout the peninsula that reveal a clear pattern: cities that have been drained for agriculture or development show steady rainfall deficits. Fort Lauderdale's deficits, for example, begin right around the population boom of 1950. On the other hand, the relatively undeveloped shrimping community of Fernandina Beach, also on the east coast, shows a steady surplus of rainfall for the same period. Chart after chart show steep deficits correlated with intense development, surpluses for slow-growing communities. The deficits in Brooksville and other areas north of Tampa begin in the late 1970s and early 1980s, when they were drained for ranchettes. Ocala's decline begins in the early 1990s, when the region's thoroughbred horse farms began to be plowed under for subdivisions. In contrast, slow-growing towns in the panhandle such as Madison and Monticello see a slight surplus. The surplus rainfall pattern would have held statewide, says O'Brien, had it not been for land-use changes.[7]

THE RISING SEAS

Local climate change is different from global warming, the gradual rise in the Earth's average surface temperature. But in the politically charged debate over global warming, Pielke was branded a skeptic because of his assertions that changes in the land, like wetlands drainage, deforestation, and urbanization, are at least as important as increases in atmospheric greenhouse gases. Pielke says he is misunderstood. "Of course the climate is changing," he says. "I'm simply not convinced that carbon dioxide is the worst problem."[8]

In the weird political landscape of early twenty-first-century America, scientists were enormously frustrated that global warming was becoming, like evolution, a matter of public opinion. "People decide whether they believe it or not . . . in spite of the scientific evidence," complained Judith Curry, chairwoman of Georgia Tech's School of Earth and Atmospheric Sciences.[9]

From the National Academy of Sciences, the fact is this: average global surface temperature has climbed about one degree Fahrenheit in the past century, with accelerated warming over the past two decades. The evidence is increasingly strong that people are to blame for most of the recent warming. Pollution from smokestacks and tailpipes changed the chemical composition of the atmosphere by building up heat-trapping greenhouse gases—primarily carbon dioxide, methane, and nitrous oxide. As a result, in North America over the next century or so, we will see warmer temperatures, increased drought, and sea-level rise.[10]

The big question for Florida and the eastern seaboard, of course, is this: just how high will the oceans creep? William H. Schlesinger is a professor of biogeochemistry and dean of the Nicholas School of the Environment and Earth Science at Duke University in North Carolina. But lately, he's been riding the Sunday morning circuits like a traveling preacher. Church congregations around the Tar Heel State have invited Schlesinger to speak about global warming and its impacts on the southeastern United States. The audiences are self-selected, he admits. "I think those who believe humans should subdue the Earth leave after service and don't stick around for my lecture," he says.[11] At the national level, evangelical Christians between 2004 and 2006 were sharply divided between the "subdue the Earth" set and "What Would Jesus Drive?" types. The National Association of Evangelicals, the nation's largest evangelical organization, sparred over but ultimately could not agree on a position statement that mankind has "a sacred responsibility to steward the Earth and not a license to abuse the creation of which we are a part."[12]

But the Christians who stuck around for Schlesinger's lectures were keenly interested in what he had to say, particularly when they saw his maps that show North Carolina's Outer Banks inundated with an 18-inch sea-level rise. Most models predict between a one-foot and three-foot sea-level rise along the eastern seaboard by 2100. Schlesinger plays it safe with the median projections, about 18 inches. That rate would hasten the demise of freshwater supplies in coastal cities vulnerable to saltwater intrusion and cause severe erosion, flooding, and destruction of wetlands—particularly in Louisiana and Florida.

That is nothing compared to what will happen if, say, Greenland's massive ice sheet melts. A team of scientists at the University of Arizona's Institute for the Study of Planet Earth reported in 2005 that warmer temperatures are melting the Arctic ice sheet and glaciers so quickly they could result in ice-free Arctic summers, not seen for a million years. "What really makes the Arctic different from the rest of the non-polar world is the permanent ice in the ground, in the ocean and on land," says the team's leader, geoscientist Jonathan T. Overpeck. "We see all of that ice melting already, and we envision that it will melt back much more dramatically in the future."[13]

If Greenland melts, it would raise sea levels by 7 meters, or 23 feet. Even a partial melting would cause a 1-meter, or 3-foot, rise.[14] Overpeck and his colleagues have used computer models to create a series of maps showing how the most vulnerable parts of the globe, including Florida, would look in the case of a 1-meter to 6-meter rise. The Florida maps show the entire coastline swamped under the 1-meter scenario. With a 6-

meter rise, all the cities along the southeast Florida coast would be covered in water, as would the Fort Myers/Naples area, St. Petersburg, Jacksonville, and low-lying parts of Orlando.

In 2005, Florida's regional planning councils began to work on coastal maps to figure out what parts of the state should begin to mitigate the impacts of sea-level rise. You cannot help but think Florida has at least one thing going for it: the state's developers are very, very good at getting rid of water. And building atop it. They have been doing so for more than 150 years.

Scientists are more and more certain that global warming will cause the sea to rise. In a different debate swirling around global warming, they are far less so.

HURRICANE ALLEY

In the wake of devastating Hurricane Katrina in August 2005, scientists, journalists, and everyone else seemed to be searching for answers to the question of whether global warming was making weather more severe. Screamed the cover of *Time* magazine that October: ARE WE MAKING HURRICANES WORSE? It makes sense. Warmer air can translate into warmer oceans—and warm oceans drive the force of hurricanes.[15]

Curry and her Georgia Tech colleagues published a study in the journal *Science* that surveyed global hurricane frequency and intensity since 1970. Theirs (and other studies concur) found no increase in the frequency of hurricanes. But they found a sustained increase in hurricane intensity. Overall, they found that the number of Category 1, 2, and 3 storms had fallen slightly, while the number of Category 4 and 5 storms had climbed dramatically. In the 1970s, the number of 4's and 5's each year averaged ten. Since 1990, the annual number nearly doubled, to eighteen. Overall, the big storms jumped from 20 percent to 35 percent of the total worldwide.[16]

Other hurricane scientists argue it is not enough to go back to 1970, the year satellites first began collecting atmospheric data and images. Hurricane severity ebbs and flows over decades. Consider Florida's bad luck in the 1920s and 1930s. Major hurricanes in 1926, 1928, and 1935 killed 2,487 people in South Florida at minimum. The 1935 hurricane, a Category 5 storm, holds the record as the most intense ever to strike the United States. Among the more than 400 dead were 256 impoverished war veterans who had been sent to the Keys by the Federal Emergency Relief Administration to fill gaps in the highway between the mainland and Key West.[17] Testifying before Congress later in 1935 on providing compensation to survivors, Florida Congressman J. Mark Wilcox admitted that while thousands had died in Florida hurricanes in the prior ten years alone, "it is not a good thing to talk about, it is a thing I hate to mention, because it is not good advertising."[18]

As it often does, nature gave Floridians plenty of time to forget. It was a quarter century before the next catastrophic hurricane hit the state. Donna, in 1960, hit the Keys, slammed into southwest Florida, barreled across the interior, then exited the northeast corner of the peninsula before traveling up the entire eastern seaboard. Donna remains the only storm of record to send hurricane-force winds into Florida, the mid-Atlantic states, and New England. The storm left houses piled like Tinkertoys in the Keys, a result of 128-mile-per-hour sustained winds and 13-foot storm surges. In all, Donna killed 50 people and caused $387 million in damage in the United States.[19]

In 1969, Florida was spared the destruction of Camille, a Category 5 hurricane, when it made landfall along the Mississippi coastline. The second-most-intense hurricane to hit the nation, Camille's wind speeds will never be known because it destroyed every wind-recording instrument on the Mississippi coast. Columbia, a small town 75 miles inland, reported 125-mile-per-hour sustained winds. Camille killed 256 people and caused $1.42 billion in damage.

And then, just like before, there was a decades-long lull in severe storms in the United States.[20] From 1969 to 1989, when Hurricane Hugo hit just north of Charleston, South Carolina, Floridians paid little heed to the possibility of hurricanes. During those two decades, the state saw another extraordinary building boom, this time along both coasts, as did other hurricane-prone areas including coastal Texas; Savannah, Georgia; Charleston, South Carolina; and Mobile, Alabama.

In 1992, Hurricane Andrew killed 25 people and caused $30 billion in property damage in South Florida. But even the costliest hurricane in U.S. history did not slow migration to the coasts. By 2005, 13 million of Florida's 17 million residents lived in coastal counties. It was a national trend. At the turn of the twenty-first century, 53 percent of Americans, or 148.3 million souls, were living in coastal counties.[21] The math looks like this: more people in harm's way, minus the wetlands and other natural barriers between them and the storms, plus more Category 4 and 5 storms, equals more casualties and costlier damages.

The federal government helps increase these odds, in a number of ways. First, the Army Corps, which regulates impacts to wetlands, permits the construction. Then, as homes and condos replace wetlands, the shoreline begins to erode. No worries. Congress gives Florida, Texas, New Jersey, and other states millions of dollars a year to rebuild their beaches. Between 2004 and 2006, local, state, and federal dollars devoted to beach and dune rebuilding in Florida alone added up to $409 million, the largest such effort ever.[22] Beach rebuilding is now the fastest-growing part of the Army Corps' work.[23] Sometimes, state and local governments come back within just a few years to ask Congress for more money after storms wash rebuilt beaches away.[24]

But the biggest subsidy for coastal living is the federal flood insurance program. Most people could not afford to live near the water without the taxpayer-funded program, which provides relatively cheap insurance—as little as $400 a year per $100,000 in coverage—in areas where private companies will not write policies.

America's approach to development and flood control—the wetlands drainage, the dams, the levees, the river channelizations, and the manipulation of the coasts—has proven foolhardy again and again. Flooding may be the worst problem faced by residents in coastal counties, but fouling up the freshwater supply is a close second. In Florida, peninsular Pinellas County was the first county in the state to blow out its natural water supply, a result of overpumping and dense development. In North Carolina, fifteen coastal counties have dried up their once-bountiful aquifer. The Coastal Plain aquifers of southern New Jersey have seen water levels drop by 200 feet in areas, with some dipping below sea level.

But nothing so painfully demonstrated the need to stop manipulating water and wetlands than the hurricane that drowned the beautiful city of New Orleans in 2005.

When Hurricane Katrina churned into Louisiana and Mississippi that fall, killing more than 1,300 people, Americans stayed glued to round-the-clock television and Internet news, horrified that one of the great cities of the United States could be extinguished like some modern-day Pompeii. The floating bodies and impoverished survivors caught twenty-first-century Americans by surprise. Some called it an act of God. Hurricane Katrina, said President George W. Bush, was "one of the worst natural disasters in our nation's history."[25]

But the catastrophe was, to a large extent, man-made. If building New Orleans below sea level was foolish, removing all natural barriers between the city and the sea was idiotic.

Louisiana and Florida, with their salty grass flats and their great tupelo swamps, have more wetlands than any other of the lower forty-eight states. But they should have a lot more than they do. Louisiana's wetland acreage has shrunk from 52 percent to 27 percent of the state, with a total 7 million acres wiped out. Florida's wetland acreage has shrunk from 54 percent to 31 percent of the state's total surface area. That includes the destruction of the mother of all wetlands—the Everglades—as well as smaller but no less significant losses that add up to more than what was lost in the great marsh. Across Florida, 9 million acres of wetlands have vanished.[26]

The benefits of wetlands to wildlife are well-known. They are the baby nurseries for most fish, homes to countless animals and birds. The ivory-billed woodpecker, believed extinct for sixty years but recently spotted in an Arkansas wildlife refuge, lived on insects in giant floodplain trees that were logged out of America's bottomlands. The extinct Carolina parakeet ate the seeds in cypress tree balls.[27] More life teems in one acre of healthy wetland than in an acre of almost any other type of habitat.[28]

But "natural disasters" such as Katrina remind us that wetlands are important to people, too. When they lost all those acres, Floridians and Louisianans lost their most important tool for flood control. They lost their best weapon against shore erosion, against fires. They got rid of a key cog in their drinking-water supply system, since wetlands absorb water during wet seasons and gradually release it during dry times. They eliminated their best natural filter for cleaning water. Because wetlands take in water from higher ground, they act as natural filters that absorb nutrients, toxic substances, and disease. A good example of the natural filtering capacity of wetlands is a cypress-gum swamp west of Tallahassee that absorbed lead from years of washing car batteries with acid, preventing serious health hazards downstream.[29]

Since the early twentieth century, Americans had destroyed more than a million acres of Louisiana's coastal wetlands, or 1,900 square miles. Between 1890 and 1925, loggers wiped out virtually all the state's virgin bald cypress forests. For the rest of the century, developers, oilmen, and the Army Corps razed tupelo and cypress stands and wiped barrier islands clean of mangroves as they cleared the path for industry, oil pipelines, levees, and canals. The wetlands and the islands, the mangroves and the trees, all helped protect coastal Louisiana by absorbing energy and water from storm surges—the most dangerous element of hurricanes.[30]

A powerful surge can swamp inland areas 20 feet or more. Scientists say that each square mile of wetlands absorbs enough water to knock the surge down by one foot.[31]

"God is getting a bum rap," environmental historian Theodore Steinberg complained about the characterization of Katrina as natural disaster. The hurricane's deadly destruction was "an unnatural disaster if there ever was one," he says, "not an act of God."[32]

No scientist could say with certainty whether Katrina's deadly force had increased due to warmer ocean temperatures caused by global climate change. What had increased, with certainty, was Americans' willingness to destroy the wetlands that mitigate the dangers of coastal living. And to build homes and businesses where they should not.

PUTTING THE WATER BACK

When the killer hurricane of September 1926 raged through Florida's Everglades, it devastated a sprawling sugar plantation northwest of Miami owned by the Pennsylvania Sugar Company. Like so many other South Florida interests during those brutal years, Pennsylvania Sugar could not withstand the double blow of first the storm and then the Great Depression. In 1931, company officials decided to write off the Everglades and Pennsuco, a company town in the swamp. As part of a severance package, they gave 7,000 acres to their Pennsuco foreman, Ernest "Cap" Graham, for the value of the outstanding taxes.[33]

Cap Graham converted to cattle, the business of his family back in Michigan. Graham Dairy became one of the best-known businesses in Dade County. Those acres brought him fortunes he could not have achieved as a sugar foreman: he made millions, served in the state senate, sent sons to Harvard. His oldest, Philip, would marry the daughter of the publisher of the *Washington Post* and make a small-time paper into a publishing empire before taking his own life. Graham's two other sons would go into that most classic Florida profession: development. They would turn their father's swampy land into the planned community of Miami Lakes, multiplying the family millions.[34]

But the youngest, Bob Graham, also was drawn to politics. His brother Philip was an inspiration. A political savant, Philip was the adviser who cajoled Jack Kennedy in 1959 to tap Texas Senator Lyndon Johnson as his running mate in order to beat Richard Nixon in the conservative South. Bob Graham would win votes for nearly half a century: four terms in the Florida legislature, two as governor of Florida, and three more in the U.S. Senate before a brief run for the U.S. presidency in 2004. (In the end, Graham withdrew, deciding to maintain his track record of never losing an election.)

Graham's biographer, *Palm Beach Post* reporter S. V. Dáte, points out that the Graham family wealth is based on the three things most damaging to Florida's Everglades: sugar farming, cattle ranching, and urbanization. Despite this, or perhaps, Dáte posits, because of it, Graham became the first governor of Florida to champion the huge task of restoring the ecosystem.[35] In other words, he wanted to put the water back.

As a young state senator from fast-growing Dade County, Graham had carried the landmark land and water bills for Askew in the upper chamber of the legislature in 1972. During his turn as governor from 1979 to 1987, Florida would see yet another round of environmental protection and development laws, including a tough growth-management law that was the first of its kind passed in a large, urbanized state.

Graham had learned the importance of well-planned growth during his family's development of Miami Lakes. (The town, it must be said, is in a drained Everglades wetland.) The charming "new town" had its own downtown, winding streams and curving streets, public parks, and underground utility lines—popular ideas now but unheard of in Florida in the 1950s. Key to the design was the company's ability to control development. "As the sole owner of all the surrounding lands, the Grahams carefully plotted where each pocket of stores and office buildings went, eliminating stretches of strip malls and industrial parks that define many faceless suburbs," wrote the *Miami Herald.* The investment paid off, as Graham Companies earned a fortune selling homes to eager suburbanites in the 1960s and 1970s.[36]

Florida's heralded 1972 laws included a planning regimen called "Developments of Regional Impact" that developers had to go through any time they built a project with larger-than-local impact. It was supposed to encourage well-planned developments like Miami Lakes. But by the time Graham moved into the governor's mansion, it was doing just the opposite. Developers complained it was expensive, time-consuming, and duplicative. So they simply did smaller projects to avoid it.[37] Piecemeal construction was the order of the day, and it showed. Florida's cities at midcentury had character. By 1980, local color was rapidly giving way to generic strip malls and low-density residential tracts.[38] That pattern, in turn, created an infrastructure crisis. The water systems were overtaxed. Traffic jams were getting longer. Schools were crowded, the sewage systems not expanding quickly enough for the 893 new people who showed up in Florida each day.[39]

Graham took office intent on changing Florida's environmental "value system." Despite a growing sensitivity toward the land, too many Floridians, he said, still saw nature as a commodity: "the purpose of Florida was to alter that commodity to make it more marketable and sell it as fast as we could," he said. Graham wanted to "convince people that we had a different relationship to our environment. That we were trustees, not consumers, and we ought to be thinking about how we can fulfill our trust responsibility."[40]

But like every other governor in Florida history, Graham worked just as hard to draw more and more people and businesses to the state, exacerbating the problem he was trying to solve. Graham pushed major infrastructure upgrades to the ports, airports, and highways to promote growth and development. In fact, Graham, credited with bringing "growth management" to Florida, never liked the term. He said it "makes it sound as if growth is something that would be bad for Florida."[41]

Graham's style was deliberative, and some called him "Governor Jello" for not moving fast enough for the fast-growing state.[42] But his environmental initiatives would prove enduring. His new growth-management law required every local government in Florida to draw up land-use plans that also had to be approved by the state. (The law does not work too well because developers are often able to convince local governments to change the plans. Especially to extend urban growth boundaries into wetlands or other boondocks.)

Most importantly, Graham set the state and federal government on their current path to restore the Everglades, in as much as it can be restored with 7 million people living in South Florida. In 1983, he unveiled his "Save Our Everglades" program, a major effort to see that "the Everglades of 2000 looks and functions more like it did in 1900 than it does today." That did not come to pass. But an important part of the restoration Graham championed has. The $500 million Kissimmee River project offers today's best hope that restoration plans for Louisiana, the Chesapeake Bay, and a great marshland in southern Iraq, drained by huge channels and canals on the order of Saddam Hussein in the early 1990s, are more than pipe dreams.

It was a crusading Fish and Wildlife Service biologist named Arthur R. Marshall Jr. who convinced Bob Graham the only way to restore the Kissimmee River was to get rid of the works of man. Marshall had a plan to blow up the engineers' dams, backfill the ramrod-straight C-38, and buy back the surrounding former wetlands that had been filled for cattle pastures. "It is time—well past time—that we abandoned the centuries-old belief that man's dominion over the earth includes its willful destruction," Marshall once said. Graham agreed.[43]

It was nearly two decades from the day Graham's Save Our Everglades plan set out to "re-establish the values of the Kissimmee River" until the first dam above Lake Okeechobee exploded into gray smoke and chunks of concrete. But explode it did. The river's restoration remained Graham's top environmental priority after he was elected to the U.S. Senate. In 1990, he slipped language into a public-works bill that authorized the Army Corps to take on purely environmental projects. *Washington Post* reporter Michael Grunwald, a historian of the Everglades who has written extensively about the Corps, called it "a little-noticed turning point for an agency that has traditionally taken on environmentally disastrous projects."[44]

On a bright, clear day in June 2000, a South Florida Water Management District scientist named Lou Toth stamped his foot on a detonator and blew up one of five dams that held back the Kissimmee River. The district and the Army Corps also removed a boat lock and backfilled seven miles of the C-38. Since then, progress on the project has slowed; large portions of the river are still channelized and dammed, and historic hydrologic conditions have yet to be fully reestablished in the recovering wetlands.[45]

But one look at the Kissimmee today leaves little doubt that humans really can fix their mistakes in the natural world: fourteen miles of nat-

The C-38 canal and one of five dams that held back the Kissimmee River, the dynamite explosion, and the free-flowing river today.
(Photographs courtesy of Lou Toth, South Florida Water Management District.)

ural, meandering river overflows the Kissimmee's banks into 10,000 acres of wetlands once drained dry. Scientists see sandbars, sandy bottom, and other signs of improvement to fish and wildlife habitat in the channel. Isolated parts of the remnant river have been reconnected and flow again. Wetland vegetation has reappeared and thrives on the riverbanks and adjacent floodplain lands. Waterfowl and wading birds, whose numbers dropped 95 percent in South Florida in the twentieth century, are flocking back to the area.[46]

A much-lesser-known law that passed during Graham's days in the governor's office probably did more than any other before or since to protect what remained of Florida's wetlands. The Warren S. Henderson Wetlands Protection Act of 1984 required, for the first time, that any dredging or filling of wetlands, including upland and isolated ones, must be permitted. Just as the Randell Act in 1967 was named for its fishing-enthusiast sponsor, so the Henderson Act was named for its sponsor, a Republican senator from Venice who also was concerned about adverse impacts to fish and wildlife. Under the law, among the many factors the state's Department of Environmental Protection must weigh before letting an applicant destroy a wetland is whether the project will harm fishing and recreation or marine productivity.[47]

Lawmakers named the act after Henderson when they realized it would be his last good deed in the Florida legislature. Certainly, they didn't do it as a reward for his behavior on a bus trip to the St. Marks Wildlife Refuge south of Tallahassee in early 1984 to look at some wetlands. Unfortunately for Henderson, a couple reporters were along for the trip, during which the senators aboard made the driver stop for booze. By the time they got to St. Marks, they were hammered. Henderson, in addition to making some crude sexist remarks, scooped up a young *Orlando Sentinel* reporter named Donna Blanton and twirled her around with glee.[48] (Blanton is now a powerful lawyer who successfully represented former secretary of state Katherine Harris in Florida's notorious presidential recount of 2000.)

The tale says a lot about the Tallahassee press corps, not infrequently a stepping stone to a career in political public relations. None of the reporters on the bus came back and filed a story about drunken lawmakers on a taxpayer-financed fieldtrip. Brian Crowley of the *Palm Beach Post,* who had not been on the trip, heard the rumors a few days later and jumped on the story. He started making the rounds of the shabby three-story press building near the capitol to interview eyewitnesses. Realizing they were about to be scooped, everyone decided to write about the scandal.

The much greater scandal, of course, was the way Florida's growth industry continued to get around the Henderson law and the many other efforts to protect wetlands that followed it.

"THE 'BURB THAT ATE THE WETLANDS"

If ever there was a town that should not exist, it is Weston, Florida. The master-planned community juts "like a thumb sticking in the eye of the Everglades,"[49] in the words of one environmentalist, in western Broward County on the state's jam-packed southeast coast. To get there, you can take Florida's Sawgrass Expressway, which wiped out sawgrass, or the Panther Parkway, which devastated panther habitat. The city of 65,000 is flush up against the remaining Everglades. If you stand at its edge with your back to the city, the vista is sawgrass, as far as you can see. Turn around, and the view is clay-colored rooftops, as far as you can see.

In the 1950s, Arthur Vining Davis, the multimillionaire cofounder of Alcoa, the Aluminum Company of America, bought 10,000 acres of swamp, sawgrass, and ranchland in southwest Broward County for an average $300 an acre. Four years before his death in 1962, Davis founded a land-development firm called Arvida. In 1974, the company announced plans to build a city of 20,000 homes in a place that was more water than land.

The Broward County Planning Council warned that the project would threaten South Florida's natural water supply, noting the site was "on the Biscayne Aquifer, the sole source of drinking water in Broward County, which lies only a few feet below the surface." The report also said that Weston was "located within a very hazardous flood plain . . . more prone to frequent and severe flooding than any other portion of the county."[50]

But Arvida had a familiar plan to make sure its new town—then called Indian Trace—would stay high and dry. The *Fort Lauderdale Sun-Sentinel* explained it in a Sunday front-page story called "The 'Burb that Ate the Wetlands": using tax-free bonds that its residents would have to pay back over the next three decades, Arvida would drain and divert the

Like most cities in southeast Florida, Weston was built on a swamp. In the background: the Everglades.
(Photograph by Rob C. Witzel, courtesy of *The Gainesville Sun.*)

water and use the dredged-up fill to raise the surrounding land. The company would scrape away tons of muck soil from the swamp, dynamite and dredge a vast system of canals and lakes at least 30 feet deep, and use the fill to make instant real estate.[51]

The county opposed the plan; commissioners passed a tough new land-use ordinance in 1977 to try to stop it. Arvida hired an army of lawyers, lobbyists, and publicists to change local politicians' minds. Even the Army Corps opposed the plan; in 1987, the Corps declared more than a third of the 10,000 acres endangered wetlands. Going over the heads of local Corps regulators, Arvida officials flew to Jacksonville to convince brass to soften the agency's stance. Even Broward County's delegation in the Florida legislature, as well as the Florida Department of Community Affairs, opposed the special tax district, which, they said, amounted to taxation without representation. But with the help of the lawyers and the lobbyists, that passed too.[52]

Today, Weston is one of the hottest addresses in South Florida.

When Hurricane Andrew in 1992 blew Miamians north with insurance checks in hand, many who could afford it settled in one of Weston's meticulous homes. Upheaval in Latin America sent swarms of wealthy, family-oriented Latinos to Weston, where they felt safe behind its many guard gates. The town's crime rate is a third of that of Broward and Palm Beach counties; a sixth that of Miami-Dade's. It is home to celebrities such as Miami Dolphins quarterback Dan Marino. It has some of the best schools in southeast Florida. Arvida officials point out that to complete their final phase of 941 acres, they had to restore and preserve another 1,559 acres of wetlands.[53]

But Weston juts like a thumb into the path of Everglades restoration, too. Weston and the development it drew to the surrounding swamplands also presents an enormous challenge for the water managers who control the Everglades' pumping stations and hundreds of miles of canals that keep Broward County's cities from flooding. The vast majority of freshwater wasted in South Florida is not really from lawn sprinklers. Rather, it is the amount drained off each day to keep the whole place dry.

More than anything, Weston opened up western Broward County for development. The town's political clout helped bring Interstate 595 to its front door, which in turn drew thousands more housing developments to the edge of the Everglades. Just like always, it was not the laws on the books that allowed thousands more acres of wetlands filled, brought thousands more people to settle in a fragile land. It was politics.

5

Red State, Green State

On a freezing morning in Jackson Hole, Wyoming, in 1971, President Richard Nixon stepped aboard a boat to take in the towering beauty of the Grand Teton Mountains—or at least, to make the American public think he could enjoy such a boat tour and the specter of the snowy, jagged range.

Late in the afternoon on the previous day, Nixon told his interior secretary, Rogers Morton, that he wanted an early-morning photo session out on the lake. It was, they had learned, the most photographic time of day, when the sun gave the signature 13,770-foot Grand Teton a golden hue.

The problem, Morton and his staff knew, was that morning temperatures were forecast in the twenties. Nixon disliked, and did not own, rugged outdoor garb. He was sure to insist on his ubiquitous shiny blue suit.

Nathaniel Reed was Morton's number two man at Interior. Reed went to work hunting sweaters, a great big coat, and a hat. He got the garments to the boat for Nixon's advance team, where the Secret Service searched them. The advance man made Reed get down in the bottom of the boat. The only two people who should be seen by the press, he said, were the president and Morton. Reed snuggled in with all the warm clothes. Nixon arrived in the shiny blue suit, a blue tie, and white dress shirt with no undershirt. He refused to don anything more.

As they chugged away from the dock with a press boat on each side, Nixon pointed to the Grand Teton. "It is beautiful!" he exclaimed.

And in a whisper, "Nathaniel, it is cold as hell."

About four minutes out, another whisper: "I am freezing my ass off. Get the hell out of here."

"The president has a very important telephone call," a Secret Service agent blared through a microphone. "The photo session is now over."[1]

Nixon, the president, is credited with the most significant environmental policies in U.S. history. He understood that Americans wanted land preservation, clean water and air. And he wanted the legacy of giving these to them. The year before his Teton tour, he created the Environmental Protection Agency, declaring it "a coordinated attack on the pollutants which debase the air we breathe, the water we drink and the land that grows our food."[2] In the years that followed, he would sign into law the Clean Water Act, the Clean Air Act, and the Endangered Species Act. All were acts of political pragmatism, tools to associate himself, and the GOP, with environmental concern.[3]

Nixon, the man, cared little for nature—and he could not relate to those who did.[4] The president who famously walked California's beaches in black wingtip dress shoes could not imagine why his constituents would want wilderness areas in Alaska and the Everglades, which he helped save from a proposed jetport. He went along, partly as a way to "take the initiative away from the Democrats."[5]

The members of America's Bush dynasty are Republicans of a very different feather. The former president George H. W. Bush and the politicians among his sons, President George W. Bush and Florida Governor Jeb Bush, all love the feel of sand beneath their feet. The three men's tan faces and arms glow with their affinity for fresh air and warm sun. From California to Maine, the Bushes walk the beaches bare legged and barefooted. In addition to the family retreat in Kennebunkport, they vacation on the wide prairies of Texas, the striking blue waters of the Florida Keys. They like to fish and bike just as well as they like a round of eighteen holes.

The Bushes appreciate nature like a commercial fisherman loves the spawning grounds, like a movie director relishes a misty swamp. Nature, for them, carries a value beyond that for ecosystems or personal appreciation. It has an economic worth that can be tallied up as easily as Cabela's annual revenues ($1.4 billion for the hunting-and-fishing superstores that have become must stops in GOP political campaigns[6]). This ideological conflict may be as old as stone tools. The nature-loving Romans mined each and every part of their empire for metals. Theodore Roosevelt, the American president who championed wilderness preservation, bagged big game. One of the most beloved naturalists in U.S. history, Aldo Leopold, was a game manager who spent his days figuring out the maximum numbers of deer or quail his hunting constituents could kill.[7]

So from the Arctic National Wildlife Refuge: oil. From the national forests: timber.

But the extraction of value from nature is a bit more complicated in Florida, whose core currency is natural amenities. Whimsical palm trees, white-sand beaches (okay, so those aren't natural), and balmy weather are its stock in trade. When it comes to environmental protection, this is often a good thing. Not only the aesthetic but the economic importance of Florida's beaches and parks means a politician cannot succeed without committing to protect them. In twenty-first-century Florida, Republicans and Democrats alike oppose oil drilling off the state's Gulf coast. They champion the restoration of the Everglades. They love water and land preservation. They vote for strict pollution controls. Their constituents, including Florida's most powerful industry—home builders—often insist on it.

Of course, those environmental issues often conflict with Florida's insatiable drive to keep growing its population and industry. And then, all bets are off.

THE MAKING OF A RED STATE

In his classic study of southern politics in 1949, the political scientist V. O. Key Jr. called Florida "The Different State." Key described Florida's political structure as "an incredible mélange of amorphous factions," most fighting among themselves within the Democratic Party.[8] Two key factors set the stage for two-party competition in the 1950s and 1960s. The first, says Florida historian David Colburn, was racial politics.

Many rural Florida "Crackers," die-hard Democrats, felt betrayed by Governor LeRoy Collins's (1955–61) moderate approach to the *Brown v. Board of Education* school desegregation decision compared with his fiery southern counterparts. The racial reforms of Presidents John F. Kennedy and Lyndon Johnson further crumbled the Crackers' faith in the Democratic Party.[9]

The second factor was the 1967 reapportionment of the legislature, long ruled by the infamous North Florida "Pork Chop Gang," which kept Florida one of the least representative states in the nation. Despite enormous population growth in South Florida, the Pork Choppers who controlled reapportionment refused to give up any seats in the legislature. So a small percentage of the population in the north had the vast majority of voting power: In 1960, 12.3 percent of Florida's voters elected a majority of the state senate, 14.7 percent a majority of the house. The state's five most populous counties had more than half the population but only 14 percent of the senators.

After decades of fighting this injustice, first in the capitol and then in the courts, urban moderates finally won in U.S. district court in Miami, which in February 1967 ordered all incumbents thrown out of office. The judges called for new elections under new boundaries before the 1967 legislative session could begin. The resulting power shift was dramatic. Dade County alone "saw its delegation grow from one senator and three representatives to nine senators and twenty-two representatives."[10]

Florida politics had suddenly shifted: from north to south, from rural to urban. The new lawmakers "were more business-oriented, more people from out of state, more willing to move forward, and more willing to make this a modern state," remembered Wade L. Hopping, an adviser to Governor Kirk who worked on reapportionment and went on to become the top lawyer and lobbyist for developers in Florida.[11]

Kirk's stunning victory in 1966; Edward J. Gurney's U.S. Senate victory in 1968; and Nixon's carrying of the Sunshine State in that year's three-way race for president all foreshadowed Florida's ultimate shift from Blue to Red.[12] However, this trio's reputation in office—Nixon's arrogance, Kirk's buffoonery, Gurney's corruption—helped keep Democrats entrenched until the early years of the twenty-first century.

In 1986, Bob Martinez of Tampa became the first Republican governor elected in Florida since Kirk two decades before. But four years later,

Martinez could not fend off a challenge by a popular Democrat named Lawton Chiles. "Walkin' Lawton" was a wizened Cracker who had served in Congress for years and never lost an election. He was what southerners call a "Yellow Dog Democrat," one who thinks any Democrat—even an ugly, yellow dog—is better than a Republican.

Chiles kept a young challenger named Jeb Bush out of the governor's mansion, too, but only for a time. Bush's historic two terms in office would lead Florida to what Colburn describes as a Republican juggernaut: by 2005, the GOP controlled every major Florida office, had a substantial majority in the state senate, had a super majority in the state house, and held twenty-one of Florida's twenty-seven congressional seats.[13]

THE GREENING OF JEB BUSH

The youngest son of George H. W. and Barbara Bush, John Ellis Bush was a good fit for Florida governor in at least two ways. First, he is a developer, following in the tradition of several of his predecessors. Second, he is fluent in Spanish, comfortable spending an entire day in Little Havana without speaking a word of English. Bilingual candidates in Florida have considerable advantage over those who don't know *arroz con pollo* from *ropa vieja* (shredded beef). Cuban and other Latin Americans make up nearly 17 percent of the state's population today, and demographers predict that by 2030 one in every four Floridians will be Hispanic.[14]

Jeb Bush honed his Spanish-language skills in Mexico. When he was eighteen, he taught English there as part of an exchange program through his prep school, Phillips Academy in Andover, Massachusetts. In the town of León in Guanajuato, Bush fell in love with a girl who had been born there: Columba Garnica Gallo. He rushed through a bachelor's degree in Latin American studies at the University of Texas in Austin and brought Columba home to Texas to marry her.

Jeb and Columba Bush moved to Miami in 1980, when they were twenty-seven and twenty-six, to work on George H. W. Bush's first presidential campaign. The nomination, of course, went to Reagan. But Miami brought good fortunes to Jeb. A powerful Cuban-American supporter of his father's, Armando Codina, gave him a job in the classic South Florida profession: selling real estate. Soon, Codina added Bush's name to his commercial real estate firm and gave him 40 percent of the profits. For the next decade, Bush followed the pages of his family's patrician playbook, in the words of the *St. Petersburg Times:* "hurry up and get rich, then go into public service."[15] In 1983, Bush told a reporter he was struggling to pay his bills. Ten years later, he was a millionaire.[16] At forty, he was ready to run for governor. Only in Florida, where most everyone is from somewhere else, could a candidate as inexperienced and new to the state as Bush be taken so seriously. His name did not hurt.

In his first bid for election in 1994, Bush challenged Governor Chiles and ran as an antigovernment, conservative firebrand. He did not have an environmental platform. In fact, he singled out the state's wildly popular land-preservation program, the largest and most successful in the nation, as a likely target for budget cuts.[17]

He also did not like Florida's development regulations. Having experienced the growth-management laws from the developer's perspective, Bush believed they were hurting Florida's economy, and he said so.

Chiles, in contrast, had spent decades perfecting every Florida politician's Janus pose: campaigning for both stronger environmental protection and more growth. He was responsible for a remarkable effort called the Governor's Commission for a Sustainable South Florida that forced sworn enemies, such as sugar barons and environmentalists, to sit at the same table until they came up with a restoration plan for the Everglades. But, like each and every governor in Florida history, he worked even harder to promote the state in an effort to lure more industry and residents.

That November, Bush lost to Chiles, who famously confounded his young challenger with these words on the night of their last debate: "The old he-coon walks just before the light of day." But Bush had made an impressive showing. It was the closest Florida gubernatorial race of the twentieth century. Bush knew that with a makeover he could move into the governor's mansion.

Over the next four years, Bush would transform himself into a moderate Republican. He greened considerably. In speeches, he began to recall the great environmental legacies of Republicans, including Nixon and Teddy Roosevelt.

Bush made his second bid for governor in 1998. This time, he proposed a bold successor to Florida's land-preservation program that would cost state taxpayers $1 billion. He vowed to fight for full funding for

Everglades restoration—at that time an $8 billion bill to state and federal taxpayers. These were not the passions of an antigovernment zealot. They were the promises of a savvy politician who had figured out the value of palm trees in a state that lures 70 million tourists a year. His new, green ethic helped send Jeb Bush to Tallahassee.

THE POLITICS OF COMPROMISE

Bubbling up from the depths of the cavernous Floridan Aquifer, nine pristine springs pump 233 million gallons of freshwater a day into the Ichetucknee River, making it the clearest, the coldest, and the cleanest in all the Sunshine State. In summertime, families from across the state, and the better-informed tourists, rent giant black inner tubes on the side of State Road 27 and float the afternoon away. The first mile down the river, those spring heads form deep, turquoise pools, icy oases in the humid North Florida woods. The river flows on for five more curvy miles, under canopies of ancient cypress and hardwood trees, before it meets up with the more-famous Suwannee. Belly-down on a tube, you don't even need a dive mask to see the white-sand river bottom. There, garfish nose around bright green grasses, red-eared cooters clamber over limestone rocks and cypress logs. If you're lucky, you'll see an otter.

In 1999, one of Florida's largest road-building companies, Anderson-Columbia, wanted to build a huge cement plant near Ichetucknee Springs State Park, the popular tube-launch spot. The plant would burn coal and old tires day and night to produce 1 million tons of cement a year, releasing hundreds of tons of pollutants in the process.[18]

The company had a long history of environmental violations. And Floridians from Miami to Pensacola—many more than the usual voices in the environmental community—flooded the office of the new governor, pleading with him to deny the permit.

Five months into his first term, Governor Bush canoed the Ichetucknee River during a press event. His newly appointed environmental chief, David Struhs, paddled nearby. At a bend in the gin-clear river, they stopped at a sandy bank to hold a press conference. Bush called the river "spectacular." He told reporters that tubers all along the route had signaled thumbs-down to the cement plant. So why not just say no to the permit? "We might," Bush said.[19]

Nine days later, Bush stunned Florida's environmentalists when he

Jeb Bush canoes the Ichetucknee River early in his eight-year tenure as governor of Florida. His change of heart to allow a giant cement plant near Ichetucknee Springs State Park was one of his more controversial environmental decisions. Bush was considered "more green than not," but his commitment to industry often trumped his environmentalism.
(Photograph by Alan Campbell, courtesy of *The Gainesville Sun.*)

did just that. His office announced it was denying Anderson-Columbia's air-quality permit. The decision hinged on a state rule that requires a company to make "reasonable assurance" it will not pollute. Struhs said the company's extensive environmental violations made such assurance doubtful. "This decision should place the regulated community on notice," Struhs said. "Compliance counts."[20]

Today, Struhs is vice president for environmental affairs at International Paper, the world's largest paper company and one of Florida's biggest polluters. Anderson-Columbia's cement plant, meanwhile, looms near the Ichetucknee River: a yellow behemoth belching tons of pollutants through tall smokestacks. Anderson-Columbia got to build it after filing a lawsuit against the state. In response to the suit, Governor Bush was the swing vote on a Florida cabinet decision to negotiate a compromise with the company and give it the permit, after all. The governor said Florida got more out of Anderson-Columbia than if the state had lost the lawsuit, and he said all indications were that the state would have lost. The company admitted past sins, put $1 million into a river-protection trust, donated land to the public, and agreed to sell the state a mine the company owned at the headwaters of the Ichetucknee. (Taxpayers paid 2,000 percent more for the mine than the company had paid for it.)

Florida politicians got a lot out of the company, too. The *Miami Herald* broke the story that within two days after cementing the deal that allowed the plant, Anderson-Columbia executives and lawyers donated $190,000 to the state GOP and President George W. Bush's presidential campaign, which Governor Bush was running in Florida at the time.[21]

Bush's flip-flop on the Ichetucknee was one of the most controversial environmental decisions in his eight-year tenure as governor. It is the one that bothers him the most, too, but not for the reasons you might think. He insists that giving Anderson-Columbia its permit was in the best interests of Florida. What he regrets is that it left him with an antigreen reputation even though he carried Florida over tall environmental hurdles, cinching the multibillion-dollar federal deal to restore the Everglades, for example, and overseeing the largest land-preservation buy in state history. "It would have been politically correct to get sued and lose, which would have happened," Bush says. "We have made great progress in lowering air pollution and improving water quality, but we don't get credit because of press coverage based on symbolism rather than actual results."[22]

It is true that by the end of his eight years in office, the conservative zealot who had first run for governor of Florida with his ax sharpened to slash environmental regulations and preservation budgets had become one of the GOP's greener governors. Bush pushed an environmental agenda comparable to those of former New Jersey governor and EPA administrator Christine Todd Whitman, New York's George Pataki, Minnesota's Tim Pawlenty, and Arnold Schwarzenegger in California. These governors tried to achieve goals such as cleaner air and water and decreasing wetlands losses in ways that, as Whitman described it, "didn't rely on the heavy hand of government but would instead build partnerships around shared goals for a better environment."[23]

So was the environment better off as a result of these partnerships—the politics of compromise that marked environmental policy in Florida and the nation in the late twentieth and early twenty-first centuries? The answer will not be clear for a decade or maybe two. That is when, at least in the case of Florida, citizens will be able to judge three things: (1) whether the plan to restore the Florida Everglades was really as much about environmental protection as it was about increasing water supply for southeast Florida's booming population, (2) whether a shift in land-preservation buys toward conservation easements that relied on private property owners to conserve sensitive areas led to stewardship or swindle, and (3) whether compromises that let developers make up for their destruction of wetlands in new ways resulted in the national goal of "no net loss" of wetlands set by Jeb Bush's father, George H. W. Bush, when he was president. The jury is still out on numbers 1 and 2.

As for number 3, it was looking as bad . . . well, as bad as a bulldozed swamp.

THE POLITICS OF WETLANDS

Florida has lost more wetlands cover than any other state, with 9.3 million acres filled in or paved over by the late 1980s.[24] Every law put in place to try to stave off draining and filling has failed—squeezed between massive population growth and political pressure from the very lawmakers who passed the statutes in the first place. After Congress passed the Clean Water Act in 1972, federal and state agencies came up with requirements for "mitigation"; that is, developers had to create new wetlands if they paved others over. But regulators frequently did not enforce the law. Meanwhile three-fourths of the artificial wetlands failed. And many of the mitigation projects initially thought successful were later found to be ecologically worthless because they were so small and isolated.[25]

By 1989, disappearing wetlands had become a crisis nationwide, one that former president George H. W. Bush vowed to reverse with a new wetlands policy called "No Net Loss." Bill Clinton and George W. Bush carried on the goal. It, too, is a failure.

Just before she left office in early 2006, President Bush's interior secretary, Gale Norton, released a report that told a surprising turn-around tale for the nation's wetlands. Between 1998 and 2004, she announced, the United States gained 191,750 acres of wetlands. "For the first time since we began to collect data in 1954," Norton said, "wetland gains have outdistanced wetland losses."[26] The report counted wetland-mitigation projects. More and more studies show that these constructed wetlands do not come close to matching real wetlands in function or value.[27] Astoundingly, the report also counted the following as wetlands: ornamental lakes in residential developments, storm-water retention ponds, wastewater treatment lagoons, aquaculture ponds, and golf course water hazards. Only by digging into the 112-page report could a reader figure out that actually more than half a million acres of naturally occurring wetlands had been lost in the six years covered.[28]

According to the study, the U.S. regions that experienced the greatest losses were the Atlantic and Gulf coastal plains. Losses also were significant in the Great Lakes states and in rapidly developing urban areas. Indeed, a regional study of southeastern Virginia between 1994 and 2000 showed a net loss of nearly 2,100 acres of wetlands, a 1.3 percent decline in wetlands in just six years. In southern Michigan, a joint report by Ducks Unlimited and the Fish and Wildlife Service documented a net loss of 30,311 acres between 1978 and 1998.[29]

As these numbers indicate, No Net Loss is a shell game. And nowhere is the game as popular, or as devastating, as in Florida. A 2005 investigation by the *St. Petersburg Times* that used satellite imagery of Florida's land cover found that in the fifteen years since No Net Loss had been in effect, at least 84,000 additional acres of Florida wetlands had disappeared. The investigation, by environmental reporters Craig Pittman and Matthew Waite, found the Army Corps approved more permits to destroy wetlands in Florida than in any other state, and allowed a higher percentage of destruction in Florida than nationally.[30]

Between 1999 and 2003, the Corps approved more than 12,000 wetland permits for Florida. It rejected one.[31]

In recent years, the most widespread paving over of wetlands in Florida occurred on the fast-growing southwest coast, along the western edge of the Everglades. In other words, just as the Corps and the South Florida Water Management District were spending billions of dollars to fix the drainage and fill problems that destroyed the swamp to the east, both agencies were permitting the same sort of damage to the west. Even assuming perfect execution of Corps' mitigation requirements, a National Wildlife Federation study found the net loss of wetlands around Fort Myers and Naples over a recent four-year period was more than 2,700 acres, or more than 600 acres a year. Annual permitted wetlands destruction for entire states is typically much lower than the 900-plus acres a year the Corps is permitting in southwest Florida.[32]

The problem is never really the Corps. It's the politicians who pressure Corps' officials, in phone calls, in e-mails, in letters, and in person, to grant permits they should not, or to otherwise bend rules. Pittman and Waite found countless examples of powerful lawmakers from both parties who leaned on the Corps to hurry up permitting. During just one three-month period in 2004, U.S. senators and members of Congress called the Corps' Florida headquarters in Jacksonville thirty-four times and wrote thirty letters about pending applications to destroy wetlands.[33]

Simply put, Florida's developers have the political power of a Category 5 hurricane. In the early years of this century, they passed farmers as the second-biggest economic driver in the state, with a $42 billion impact, behind tourism.[34] In 2005, Florida's top three political leaders were all in the business: the governor a real estate developer, the senate president a home builder, and the house Speaker a paving contractor. It was no wonder regulators were not just bending the rules but, in the last undeveloped part of Florida, letting a major developer not even follow them.

A MILLION ACRES

Since the Great Depression, Florida's largest private landowner has been a company called St. Joe. The timber-and-paper conglomerate was built up by a shrewd financial genius named Ed Ball, who famously offered up this bourbon toast each night: "Confusion to the enemy!" Ball bought up acreage during Florida's real estate bust years in the 1920s and 1930s, a time when some settlers saw sandy beaches as useless because they were not fit for growing potatoes. He picked up the company's twelve miles of oceanfront property, for example, for less than $15 an acre.[35]

Thanks to Ball, St. Joe would come to own an unmortgaged million acres in Florida's sleepy, undeveloped panhandle. Also thanks to him, the company would wield political influence in Tallahassee wide as its massive forests of planted pine. So it made sense that, in the 1990s, the company would begin to turn from trees to towns. In 1997, St. Joe bought real estate powerhouse Arvida, developer of Weston, and a month later became partner in Miami-based Codina Group, Governor Bush's old company.[36] In short order, St. Joe became the state's most ambitious developer since Hamilton Disston.

These days, St. Joe officials call the company JOE, its symbol on the New York Stock Exchange. They are transforming a rustic, rural region the size of Long Island into upscale communities worthy of the pages of *Coastal Living* (in which they are heavily advertised). JOE has twenty developments in various stages of planning and construction in the panhandle, with permits to build more than 10,000 homes, and many more on the way.[37] The company is building hotels, hospitals, schools, golf courses, shopping centers, theaters, restaurants, offices, and industrial parks.[38] "I'll tell you what the historians will say 100 years from now," the then U.S. representative Joe Scarborough told the *St. Petersburg Times*. "They'll say that St. Joe finally put northwest Florida on Florida's map."[39]

That statement is no doubt true; whether for good or for bad remains to be seen. Like its forerunners in the nineteenth and twentieth centuries, JOE has Florida's leaders convinced that its brand of growth will benefit the panhandle, with well-planned communities that spark high-end economic development. Also like the development companies of old, JOE uses the irresistible promise of the fortunes of growth to land significant subsidies from taxpayers, such as the relocation and huge expansion of an airport on land it donated, and significant exceptions to environmental regulations. Nowhere is the last point more clear than in the case of state and federal wetland laws.

In 2004, Florida's Department of Environmental Protection signed an unprecedented agreement that freed St. Joe from standard wetlands permits. In exchange for the conservation of 20,000 acres, the Department of Environmental Protection gave St. Joe a blanket permit to fill wetlands in other parts of its holdings. Governor Bush's environmental regulators hailed the agreement as a model for regional growth that would protect a huge chunk of wetlands rather than fragments.[40] The deal illustrates the funny math inherent in No Net Loss. As Stetson University environmental law professor Royal Gardner, a former Army Corps lawyer, puts it: "If a developer fills five acres of wetland in exchange for agreeing to preserve ten acres, the immediate net result is still a loss of five acres."[41]

After initially saying no way to the St. Joe deal, Army Corps officials agreed to the variance after three years of closed-door meetings. Instead of a permit for each project, which would require public notice and input from neighbors and environmental groups, the Corps gave the company broad approval to destroy 2,000 acres of wetlands in various places.[42]

Supporters of the plan had an excellent point: the laws on the books were not working to protect Florida's wetlands, so why not try something new? It is also true that JOE is an environmentally aware development company, "intent on doing it right," in the words of CEO Peter Rummell. Because of its large, contiguous holdings, "JOE has the chance to avoid many of the problems that have plagued other parts of Florida through planning that carefully considers community and environmental impacts."[43]

Florida's environmental history is filled with intentions to get it right. The question is not Rummell's intent but whether JOE's shareholders will remain satisfied with the company's performance to see it carried out.

The problem is precisely the same when it comes to conservation easements, a relatively new tool in land preservation. When a government enters into a conservation easement, it does not buy land outright but instead purchases development rights—allowing private landowners to continue activities such as timber harvesting or cattle ranching. The easements look good today, allowing Florida, other states, and the federal government to conserve thousands of acres for far less than it would cost to buy them free and clear. But how heirs and other unknowns will shape these lands tomorrow is anybody's guess.

IN PERPETUITY?

Florida Forever, the current name of the state's land-buying effort that dates back to 1964, is widely regarded as the most successful such program in the United States. Funded by a documentary-stamp tax on real estate transactions, the program and its predecessors have spent nearly $4 billion to conserve nearly 4 million acres of the sort of land that makes Florida Florida. Over four decades, the programs have saved wildlands all over the state, from the Florida Keys ecosystem on the southernmost tip to Perdido Key in the westernmost corner.

In recent years, skyrocketing land prices and fast-shrinking numbers of large, environmentally sensitive parcels have presented the state tough choices about how to get the most for its preservation dollars. Governor Bush pushed the Department of Environmental Protection to recalibrate its acquisition strategy: to focus less on ecologically pristine jewels and more on large tracts to protect vast acreage for less money. Many of those, such as a tract called Picayune Strand in Collier County, are crucial water-storage areas. Bush also pushed the state toward conservation easements. Florida bought its first easement in 1999. By the end of 2005, it had entered into 77 of them.[44]

Supporters of easements say they are often the only way the government can protect conservation land, at half as much as it would cost to buy it outright. But the easements also limit public access to lands, an ostensibly important criterion for Florida Forever buys. And although they are written in perpetuity, "everyone is concerned about the next generation," acknowledges Richard Hilsenbeck, who helps broker preservation-land buys for Florida's chapter of the Nature Conservancy. Most of Florida's easements are with longtime farm families whose land is well managed. But in some parts of the United States, heirs have sued estates to challenge the easements. New owners or heirs also have terminated easements, knowing they could make far more from violating them than they would be penalized.[45]

Perhaps more worrisome is this: some of the easements give up so much to developers that they are beginning to look like the land giveaways that got the state into so much trouble in the nineteenth century.

In 2004, Bush and the Florida cabinet approved what may be the weirdest land deal ever in a state whose history is full of them. The $18 million deal would save Cypress Gardens, a down-on-its-luck, 1930s-vintage theme park near Winter Haven known for its champion waterskiers, hoop-skirted Southern belles, and lush botanical gardens. Flagging attendance forced the family that owned Cypress Gardens for nearly seventy years to put it on the market. To keep the park from being paved over for subdivisions, the state partnered up with a man named Kent Buescher, president of a Valdosta, Georgia, theme park called Wild Adventures. Buescher paid only $7 million to become the new owner of the 150-acre park. Using Florida Forever funds (which can be spent to save cultural resources), the state paid $11 million for a conservation easement that preserves 30 acres of gardens and lakefront and puts deed restrictions on the remaining 120 acres.

The easement may prevent rooftops, but roller coasters are okay. The deed restrictions let Buescher build a new amusement park, roller coasters and all, on the 120 acres. "Granted," said Governor Bush just before the vote to save the amusement park, "in nowhere but Florida would this be a cultural resource."[46]

To his credit, Jeb Bush also pushed through the biggest, and the most expensive, land-preservation deal in Florida's history. The 91,000-acre Babcock Ranch, straddling Charlotte and Lee counties on the fast-growing southwest coast, had been on Florida's preservation wish list for years, the final link in a 65-mile-long corridor of preserved land from Lake Okeechobee to Charlotte Harbor. The ranch is home to panthers, black bears, and wood storks, and it is an enormous water-storage area in a region whose cities are running dry.[47]

The Babcock family, well-known environmental stewards, had run the ranch since 1918. In a familiar Florida story, when the patriarch died in 1997, his heirs decided to sell. The state tried to buy the ranch outright for $455 million, but the family said no. Selling would mean too steep a tax bill. Instead, they agreed to a complex conservation easement in which they sold the ranch to a Palm Beach developer who would, in turn, sell most of the land to the state for preservation but develop 18,000 acres and use some of his profits to pay the Babcock family's taxes. Florida was able to preserve 74,000 acres for $350 million. On the remainder of the land, the developer planned a brand-new town with 19,500 homes—all in the middle of the boondocks.[48]

The increasing reliance on conservation easements and the Babcock deal were examples of the politics of compromise that marked Florida and U.S. environmental policy in the early years of the twenty-first century. The massive project to restore America's Everglades was the ultimate example.

WATER FOR MAN, WATER FOR NATURE

In November 2000, as the nation was paralyzed with the uncertainty of whether Albert Gore Jr. or George W. Bush would be moving to the White House, a smiling Jeb Bush joined Bill Clinton in the Oval Office as the outgoing president signed the bill to approve the Comprehensive Everglades Restoration Plan. It would be the largest public works project ever undertaken. The plan, passed unanimously by the Florida legislature and overwhelmingly by Congress, was a remarkable political feat considering the longtime enemies who formed a coalition to support it: from sugarcane growers to environmentalists to developers.

Two decades after Bob Graham painted the possibility of a restored Everglades and another decade after the late Lawton Chiles brought

. . . her the coalition that came up with a way to do it, Jeb Bush spent . . . ical capital convincing Congress to approve the plan. It . . . ngressional leaders and federal taxpayers might go for . . . d fix America's Everglades and develop the man- . . . oring the Chesapeake Bay, Great Lakes, and . . . e last thing Congress wanted to do was . . . the rich cities on the southeast coast of . . . id former U.S. senator Bob Smith, the . . . o chaired the Senate Natural Resources . . . e plan to his colleagues.[49] (Ranked the . . . ember of the Senate by various right-wing . . . r Clinton's impeachment and once bran- . . . Senate floor,[50] retired to Sarasota and now . . . dation, a bipartisan environmental philan- . . . ing the swamp.)

. . . that will benefit from Everglades restoration include hum . . . n aims to ensure water for nature, for farms, and for cities. Had this not been the case, the plan never would have made it out of Chiles's work group. "Really what's at stake is whether people can live in south Florida," said Lawrence Belli, deputy superintendent of Everglades National Park.[51]

The Comprehensive Everglades Restoration Plan is a complicated interlay of 68 major environmental projects over 35 years. But at its most basic, the plan aims to "get the water right." It takes back the 2 billion gallons of water a day that would flood the farms and cities of southeast Florida were it not locked up in 1,000 miles of canals, 720 miles of levees, and 200 control structures. The plan proposes to capture most of this water in hundreds of thousands of acres of restored wetlands, in huge reservoirs, and in more than 300 underground wells.

Instead of pushing the water through canals out both sides of the peninsula and into the sea, water managers will send some of it flowing into the 2 million acres that remain of the Everglades, where 69 species are endangered due to disappearing habitat. And they will send some to the cities of southeast Florida, which need 50 percent more water to accommodate projected population growth. The tension in the Everglades plan, through all those years negotiating it and probably for the next three decades carrying it out, is the balance. How much water will go to lawn sprinklers, and how much to the spoonbills?

AND THE WINNER IS . . . THE LAWN SPRINKLERS

Florida's politicians are all for the spoonbills; the whimsical pink birds represent the water and the sun and all the other natural amenities that make Florida the destination for 70 million tourists a year and the home to 17 million people and climbing. But the politics of compromise almost always put the spoonbills second. It does not matter which political party is in power. It does not matter how green the reputation of the current governor.

In the years after Congress passed the restoration plan, it began to look more and more like "a massive urban and agricultural water supply project," Richard Harvey, the EPA's South Florida director, told the *Washington Post*.[52] Two factors helped tip the scales toward water supply. First, South Florida's water crisis was growing worse by the day. Lake Okeechobee water pumped east and west to sea was causing deadly algae blooms and fish kills on both coasts. Meanwhile state environmental officials told Miami-Dade commissioners they could take no more groundwater; they had to find new water sources or stop approving new subdivisions. Second, Congress was not anteing up its half of the restoration money in time for the major projects in the plan to get started.

Three weeks before the November 2004 election that would keep his brother in the White House, Governor Bush announced an ambitious plan for Florida to borrow money and begin building some of the Everglades projects without the federal government. Called "Acceler8," the plan aimed to complete eight major projects to expand water storage, improve water quality, and restore water flows ten years ahead of schedule. It included reservoir capacity for 418,000 acre-feet of water—equal to 6 million residential swimming pools. Environmentalists worried that Acceler8 focused too heavily on urban water supply, and that it did not include projects with the most direct benefit to the Everglades, especially Everglades National Park. But without it, Bush retorted, no projects would be built at all. "We don't need their (environmentalists) permission to save the Everglades," Bush reportedly told his aides.[53]

In his book *Nixon and the Environment,* J. Brooks Flippen argues that single-minded environmental activists with unrealistic demands killed the blossoming bipartisan support for the environment in the United States. Toward the end of Nixon's first term, environmentalists criticized him bitterly, with no appreciation for what he had already achieved and no capacity to accept his business constituency. The activists "did not know how to say thank-you to Nixon," said David Brower, the former Sierra Club leader who died of cancer in 2000 after seventy years of environmental activism. The Republican president "had great promise and did great things," Brower said, "but we deserted him."[54] In response, Nixon deserted the environmentalists and sided with polluters to oppose any further environmental reforms for the remainder of his days in the White House.

In Florida, as environmentalists seemed to be losing the tug-of-war in the Everglades between water supply and restoration, they began to wonder if they should have gone along with the grand Comprehensive Everglades Restoration Plan in the first place. The issue was one of many that revealed the waning influence of environmentalists in a state whose activists had been at the national vanguard for most of the twentieth century, ever since Florida outlawed plume hunting in 1901. Particularly powerful in the 1970s, Florida's environmentalists had stopped the jetport proposed for the middle of the Everglades. They had halted construction of the Cross-Florida Barge Canal, an east-west waterway that would have cut the state in two, a dream of Florida's promoters since statehood. A group called Conservation 70s was so effective that it shepherded 41 environmental bills through the legislature in 1970 alone. Environmentalists had so much muscle in Askew's time that the massive land and water bills of 1972 passed almost entirely under developers' radar screen. "The business community was just outgunned," remembered Hopping, the developer lobbyist.[55]

It was somehow symbolic that around the turn of the new century, Florida lost its three grand dames of conservation: Marjory Stoneman Douglas in South Florida, Gloria Rains on the southwest coast, and Marjorie Harris Carr in North Florida. Just as their peers at the national level were lamenting "the death of environmentalism" after losing an all-out battle to stop President Bush's reelection,[56] Florida's next generation of environmental leaders seemed to be in a crisis of confidence.

Florida's environmentalists had succumbed to the politics of compromise in an effort to achieve their ultimate dream of saving what remained of the Everglades. Whether that decision would turn out to be a save or a sell-out remained to be seen. But it was clear that in a newly Red state, environmentalists had to start winning Republican hearts. Governor

Bush's record in Florida "has been more green than not," said Eric Draper, policy director for Audubon of Florida. "Governors are a product of their times," he said. "Even if Bush had been a great environmentalist in the vein of Graham or Askew, the legislature and the development community would have never let him get away with it."[57]

A JOB FOR THE GOP

Jeb Bush put unprecedented focus on building new water-supply projects outside the Everglades, as well—all over Florida. This is apparently in the job description for Republican governors in the Sunbelt. When he was governor of Texas, President Bush signed Senate Bill 1, the biggest change in water law since the Rule of Capture (known as the Law of the Biggest Pump) affirmed by the Texas Supreme Court in 1904 gave landowners the right to suck up unlimited amounts of groundwater even if it hurt their neighbors. The biggest pump still rules in Texas, but Senate Bill 1, carried by Senator J. E. "Buster" Brown of Lake Jackson, set up sixteen regional planning districts and charged them with coming up with water development and conservation plans to prepare for droughts and ensure future supplies for people, economic development, agriculture, and nature.

Early in his reelection year of 2006, California Governor Schwarzenegger proposed a massive public works project he called his "strategic growth plan" to upgrade everything from highways to levees. The $222.6 billion plan included money for huge new reservoirs and dam upgrades. Peter Gleick of the Pacific Institute told California's Senate Committee on Natural Resources and Water that the water-infrastructure plans were wrongheaded when the Golden State's water use was decreasing all the time, even with population growth. The proposal for new dams and reservoirs "is a serious financial, environmental and political mistake," he testified. "The same amount of money spent on reducing water waste would be far more productive." California's Department of Water Resources projected that the state's water demand will decline over the next quarter century because of efficiencies and conversion from agricultural to urban uses. Water use in California was less in 2001 than in 1975 even as the population increased 60 percent in the same period.[58]

While California had kicked the assumption that population and economic growth had to increase water demand, Florida seemed to be growing more addicted to it. Environmental regulators and water managers could point to a barrage of conservation programs. But they had not begun to lower Florida's per capita consumption, the trend in the rest of the nation.

In 2005, Governor Bush and a Republican Florida senator named Paula Dockery from Polk County, where water is increasingly scarce, led a major revision of Florida water law that put unprecedented focus on finding new water. Senate Bill 444 required Florida's regions to work together to plan long-term for water supply and build infrastructure fast, before the next drought. The new law allocated at least $100 million a year for grants to help Florida communities build up new supplies, like desalination plants. It may not have seemed conservation-oriented, says Dockery, "but what could be more conservation-oriented than getting people off groundwater?"[59]

The state's powerful home builders were insisting that Florida's leaders find them more water. And, despite water shortages, they were insisting the leaders work to lure more people down to Florida, too. Only with increasing population and new water supplies could Florida grow its most important crop: rooftops.

6

Destination: Florida

MARCH 12, 2002, was a thrilling day for Al Hoffman. An inveterate risk taker, Florida's biggest home builder took his WCI Communities public despite an economic downturn that had dried up interest in IPOs. In the morning, Hoffman watched his company sell 6.9 million shares at $19 each, raising $131 million. As the day went on, he watched the new shares rise 20 percent more before closing at $22.70.[1]

Wall Street, that spring day, was buying up Florida living in a frenzy that harkened back to the Sunshine State real estate boom of the 1920s. WCI (NYSE: WCI) is book-ending the Everglades with 36 master-planned communities on both coasts of South Florida. (Golf holes: 600. Boat slips: 1,000. Average home price: $528,000.) The company's marketing research into the 76 million strong baby boom generation now reaching retirement shows that natural amenities are a top draw. WCI's pitch: "From the tree-lined fairways of west central Florida to the beachfront condominiums of Naples and Fort Myers, to the beauty and luxury of Palm Beach and southeast Florida, WCI is building communities that embrace the native environment and create an unparalleled living experience."[2]

But as he savored his business success, Hoffman fretted privately, at least to the man he had helped send to the governor's mansion.

A former fighter pilot with a penchant for polo, Alfred Hoffman Jr. got his first taste of luxury housing as a child growing up in Chicago. His dad was a laborer in a poultry slaughterhouse on the South Side who daydreamed of investing in real estate. As a summer treat, Alfred Hoffman Sr. would buy his kids ice cream cones and drive them to the Windy City's luxury neighborhoods to ooh and aah over rich people's houses.[3] Those drives inspired Hoffman to chase his father's dreams. West Point, a Harvard MBA, and Florida's booming retiree real estate market helped him achieve them while he was still in his forties. He struck it rich developing one of Florida's megaretirement communities, 14,000-resident Sun City Center south of Tampa, in the 1970s.

Hoffman, now in his seventies, has retired as CEO of WCI Communities, the home-building behemoth he founded and took public. He is bald, a tad bit frail. But his small, ice-blue eyes twinkle with the pull he still has on elected officials from the White House down. In 2005, President Bush named him ambassador to Portugal. Campaign contributions are no small part of his sway. Hoffman and his immediate family made $434,394 in federal political contributions during the 2000, 2002, and 2004 election cycles, none of which went to Democrats.[4] As finance chairman for the Republican National Committee during both of George W. Bush's presidential campaigns, Hoffman proved he could convince others to dig deep, too. During a single tent party at his Fort Myers home in 2003, Hoffman raised $1.7 million for President Bush's reelection bid.[5] He was a similar sensation heading up fund-raising for Jeb Bush's successful gubernatorial campaigns in 1998 and 2002.

In summer 2002, after the president signed the McCain-Feingold campaign finance bill, soft money had become a politically incorrect conversation topic. But Hoffman has never been PC. He is notorious for calling the endangered Florida panther a "bastardized species" whose tiny numbers stand in the way of more aggressive development in South Florida.[6] That June, Hoffman asked a *Miami Herald* reporter at a fund-raising gala: "Who is to judge how much is obscene? How much is too much? Compared to what?" He complained about Americans spending an estimated $4 billion each year on pornography. He fretted that people spent more on Starbucks than on campaigns. "I call that obscene," he said.[7]

Turns out, Al Hoffman that June had something heavier weighing on him. An imminent threat, he thought, to WCI Communities. That month, as Jeb Bush's reelection campaign was heating up, Hoffman sent the governor a memo, talking points, and a sixty-page study commissioned by WCI: "The Impacts of Florida's Mature Residents."[8] The memo raised this alarm: the percentage of seniors migrating to Florida had declined in both the 1980 and 1990 U.S. census. Hoffman quoted Charles Longino, a national expert on senior migration trends, who predicted another drop in the 2000 census. "You can couple this negative trend with the fact that a growing percentage of Florida's seniors are leaving the state," Hoffman wrote. The developer highlighted in bold the fact that he believed would be most alarming to Bush, and a trend that might alarm WCI's new shareholders as well: "**North and South Carolina, combined, attracted more relocating retirees than did Florida in 2000.**" He concluded: "We must act before Florida loses untold billions in revenue and its sustained quality of life to other states."[9]

No one is seriously worried, of course, that Florida is going to stop growing. Or even that it will no longer be a top retirement spot. By 2015, demographers predict, Florida's population will reach 21 million, making it the third-largest state in the nation.[10] By then, it also will be the oldest state, with 5.5 million residents older than sixty-five.[11] Instead, Hoffman, Bush, and other business and political leaders fear Florida is losing its share of the wealthiest retirees: those who fill the state's coffers paying Florida's property, sales, and intangibles taxes—those who can afford luxury housing units like the ones in WCI's Lesina, a shiny condominium tower rising 20 stories above the mangrove swamps of McIlvane Bay in southwest Florida between Naples and Marco Island. The cheapest units cost $900,000.

Hoffman's report boasted that Florida's current crop of retirees generates $2.7 billion a year in sales and use taxes yet costs the state only $1.28 billion in health and human services. Advocates for the elderly argue that is because Florida scrimps on care for lower-income elders. For example, the state's waiting list for those in need of basic services, such as home health care, to keep them out of nursing homes is 15,000 and climbing. Consider the consequences if the balance were to tip: a growing lower-income retiree population and fewer top-bracket taxpayers to help cover costly elder services. It is easy to see how the specter of losing wealthy baby boomers to states like North Carolina or Arizona strikes fear into housing developers and politicians alike. And that is why

Jeb Bush jumped on Al Hoffman's problem, despite its unpopularity in an election year.

Regardless of their political stripes, most Floridians lament the state's rampant population growth as they bake in bumper-to-bumper traffic watching another orange grove make way for another drugstore chain. Many were astonished to read about Bush's proposed solution to the retiree predicament: an aggressive marketing effort called Destination Florida.

In July 2002, Governor Bush appointed a fourteen-member commission to lure more retirees to the Sunshine State. (Its honorary chairman: beloved former Miami Dolphins coach Don Shula, now a steakhouse mogul.) Bush charged the commission with figuring out how Florida could better compete for retirees with other Sunbelt havens, and particularly how it could woo those 76 million baby boomers flush with inheritance cash. "For more than half a century, Florida's beautiful natural environment and rich cultural heritage have made our state the destination of choice for retirees," the governor said in a press release touting Destination Florida. "We are committed to maintaining our first-place ranking, and will aggressively sell the advantages of living in Florida to all of our nation's seniors."[12]

The idea for Destination Florida had come from a WCI vice president named Ken Plonski. Hoffman had hired Plonski, an affable PR man, away from his top competitor, Del Webb—developer of Arizona's sprawling Sun City and other megaretirement communities throughout the Sunbelt and eastern seaboard. Seven years before, Plonski helped lead a successful marketing effort in Arizona to woo more seniors. That campaign is credited with bringing the state up to third, behind Florida and California, in luring new retirees. But there was one thing about Arizona's effort that Plonski did not mention to Bush. By 1999, Arizona taxpayers and lawmakers had begun to question seriously whether the benefits of all those new elders outweighed the health care costs and transportation problems they were creating. That year, Arizona quit funding the effort.[13]

One of the puzzling things about the Destination Florida campaign was its emphasis on marketing, on public relations, and on other states' promotional efforts—rather than a hard look at what about balmy Florida could possibly be turning retirees off. For decades, state demographers had noted a growing trend among a group of retirees known as "half backs." These are elders who retired from the Midwest or Northeast to the Sunshine State, then decided to move halfway back home: to North Carolina or Virginia—someplace where they could escape Florida's crowds and stifling summer heat. Why are a growing percentage of seniors leaving? It is not to avoid state income taxes. The rival states have them; Florida does not. It is hard to believe that they are being lured by better marketing campaigns. But that is what the governor seemed to be saying. "In the 1980s, several states in and out of the Sun Belt have formed retiree-attraction programs," read the Destination Florida press release. "As a result, Florida began to see a decline in seniors choosing to relocate and retire in the Sunshine State."[14]

Could anything else be to blame? Could it be the water problems? The teeth-clenching gridlock on Interstate 95 through South Florida, Interstate 4 through Central Florida? Could it be, for some, the cookie-cutter sprawl that makes many Florida vistas indiscernible from those in other parts of suburban America? Or school districts so crowded that one county banned children's backpacks because they took up too much space in the halls? Such questions could have started with the Destination Florida commissioners themselves. Several members, including chairman T. O'Neal Douglas, a retired Jacksonville insurance executive, had to fly into Florida for the meetings from second homes in the mountains of western North Carolina or other rival retirement states.

Bush's office launched its retiree-recruitment effort with statistics to try to help Floridians see the problem. In the 1960s and 1970s, the state's age fifty-five and older population had grown by 65 percent per decade. During the 1980s, the percentage had dropped to 29. During the 1990s, it dropped to 19.

The stats did not make the sell. Floridians—many who had called the state home for generations, many who had just moved down—were furious. A torrent of letters to the governor's office questioned the need for another classic Florida marketing campaign as they expressed concerns over pressures created by growth.

People were particularly upset to see the recruitment effort following one of the most severe droughts in Florida history. The drought had sparked more than wildfires. It brought up an old debate. From Fort Lauderdale to Jacksonville to Tampa, Floridians asked government officials the same question over and over again: why does the state approve so much new development when it does not have enough water for the residents who live here now?

Barbara Down of the Villages, one of Florida's largest retirement communities, with more than eight thousand residents living in a huge, low-density maze north of Orlando, wrote Governor Bush a letter that was typical of the response to Destination Florida. "I have moved to what I thought would be 'paradise' only to find there are critical water problems and heavy-duty sinkhole activity," Down wrote. "With regard to water, we are told to conserve—which we do—and then they use the adjusted water volumes to justify new growth."[15]

Barbara Matthews of Tampa tapped off an e-mail: "With all due respect, after having lived in this state for 30 years, I would think it would be more to the point to ensure that those of us who live here now will have enough water without being severely restricted in its use, adequate roadways without destroying even more of the environment and necessary services . . . before inviting more hordes to cross the state line and create more problems than we already have."[16]

Bill Nichols of Winter Park put it more bluntly. "We need more seniors," he wrote, "like we need more sinkholes."[17]

TURNING POINT

In the fall of 2002, as Jeb Bush's Destination Florida group met around the state to gather testimony from invited experts, the nation's drought began to wane. In New York, Mayor Michael Bloomberg lifted the city's water restrictions, and cabbies could wash their cars again. New Jersey also lifted limits on car washing and lawn watering but remained in a state of emergency because of depleted groundwater levels.[18] In Florida, the rivers began to brim again. Dry lake beds gradually filled. But it was not enough to replenish the overdrawn aquifers. In some parts of Florida, developers heard a dreaded word: moratorium. In Hillsborough County, for example, Plant City told companies including WCI that it might not be able to provide water to new developments already approved by the city commission.[19]

The drought and its aftermath represented a turning point in the politics of water in Florida. Developers had spent 150 years desperately trying to drain the state and get rid of water. Now, they were desperate to find it. Up to now, water issues had been the passion and purview of environmentalists, farmers, and the state's vast water bureaucracy created in 1972. Suddenly, water supply became the top concern of Florida's business community. Al Hoffman made sure of it.

While head of the largest community-development company in Florida as well as the Republican National Committee's finances, Hoffman also chaired the most powerful business group in Florida, the Council of 100. The private, invitation-only association of Florida CEOs was created in 1961 by a governor named Farris Bryant, who wanted a trusted panel to "provide advice to him on key Florida issues from a business perspective."[20] For more than forty years, the council had helped sway policy in Tallahassee. Recently, Bush had tapped the council to help him sell controversial overhauls of Florida's civil service and education systems.[21]

Now, Hoffman wanted the council to dive into water supply. With Bush's blessing, he appointed thirty of its CEOs to a task force that would recommend "bold action to assure an adequate water supply" for Florida.[22] They included Gary Morse, developer of the Villages, and Llywd Ecclestone, an idol in golf course community development for his PGA National Resort of Palm Beach Gardens. Hoffman put yet another developer, Lee Arnold, in charge of the task force.

A licensed commercial pilot partial to aviator sunglasses and black leather jackets, Lee Arnold is the CEO of a commercial real estate firm called Colliers Arnold, which has offices in Clearwater, Tampa, Orlando, and Fort Myers. He likes to say that he specializes in "problem properties," that is, anything that may have a problem snagging local or state development approval. As chairman of the Council of 100's water-supply task force, Arnold invited a select group of experts to advise the group, which met secretly over the course of a year. The majority of the advisers thanked in the task force's final report were lawyer-lobbyists who worked for developers. Few were scientists. None were environmentalists.[23] Arnold kept Hoffman and Bush apprised of the task force's progress. At some point, Bush must have expressed concern about the lack of input from the state's considerable environmental community. In an e-mail to Bush in the winter of 2003, Arnold shared with the governor white papers, PowerPoint presentations, and other parts of the task force's work. He wrote: "We will add now the envior [*sic*] folks as per your suggestion. Anytime you want to discuss water give me a call I will burn gas in my airplane and buy lunch." (Arnold went on to recommend Bush keep PowerBars on hand and read the book *Body for Life,* a rigorous-but-popular diet and exercise program. "You have such stress and are doing great work at the expense of your future health," Arnold wrote. "CARPE DIEM. You will be remembered for your water fix forever.")[24]

But Florida was not in for an easy water fix. Arnold was genuinely surprised when the work of his task force, made public by the *St. Petersburg Times* newspaper in the fall of 2003, churned a tidal wave of controversy. Buried in thirty-four pages of facts about projected population growth and water resources, and well-reasoned arguments about what might be done, was the recommendation that the state consider "a system that enables water distribution from water-rich areas to water-poor areas."[25] The translation was clear to anyone who had followed Florida politics for a while. The developers had drawn a bead on the state's storied Suwannee River, targeted every decade or so by schemes to pipe water from north to south. Arnold declared the Suwannee region "sitting in the Saudi Arabia of water."[26] It was an unfortunate metaphor, given that Saudi Arabia has some of the worst water scarcity on the planet.

Winding 250 miles from the Okefenokee Swamp in southeastern Georgia to the Gulf of Mexico, the Suwannee is the color of long-steeped iced tea, as inviting as a glass of it on a Florida summer day. Along its jungle-covered banks, Spanish moss hangs from above. Bumpy cypress knees protrude from below. Several of the river's bends open to deep, azure springs; some secret, some packed with buzz-cut local boys and Speedo-clad Europeans on hot afternoons. Black birds called anhingas sun themselves on half-submerged logs. Great blue herons fish along the sandy shore. And every once in a while, a giant, prehistoric fish called a Gulf sturgeon will fling itself out of the river and land with a splash. All of which gives the river a magical feel, like being transported into a novel by J. R. R. Tolkien.

Born-again believers from small towns near the river, along with lots of babies each year, are still baptized in the Suwannee. But people from throughout the South take pride in the river that is the least polluted, and the least obstructed, of all major rivers in the United States. Though he never saw the Suwannee, it inspired nineteenth-century songwriter Stephen Foster, of "Oh! Susanna" and "Camptown Races" fame, to write what would become Florida's state song. Even in the southern reaches of the state, Floridians were incensed over the Council of 100's suggestions that North Florida's water resources might be diverted to help fuel South Florida's growth.

The developers might as well have recommended that the state change its official song from "Way Down Upon the Suwannee River" to Grandmaster Flash's hip-hop version of "New York, New York."

"OUR WATER IS NOT FOR SALE!"

Some saw the Council of 100's report as the opening shot in a brand-new water war for Florida. Over the years, communities from Miami to Jacksonville had, during drought, battled over rights to water. None of those conflicts were as painful, or long ranging, as the aptly named Tampa Bay Water Wars.

The fights over freshwater in the counties surrounding wide-mouthed Tampa Bay on the west coast of Florida have roots in the 1920s, when Hillsborough County's coastal leaders split off and created their own county, called Pinellas. Just like Florida, Pinellas County is a peninsula surrounded by water on three sides. It is home to some of the loveliest white-sand beaches in all of Florida, and to some of the most spectacular art deco resorts in the state as well: the pink Renaissance Vinoy on Tampa Bay, the even-pinker Don CeSar on the Gulf of Mexico. Also just like Florida, the county is so beautiful that it is being loved to death. It is the densest of all the state's 67 counties, with 3,392 people packed into every square mile. Florida averages 296 people for each square mile of land. The nationwide average is 80.[27]

With only four bridges between Pinellas and the mainland, it is not the place you would want to be if a major hurricane came ashore at Tampa Bay. In that way it is like New Orleans used to be: a disaster waiting to happen.

Pinellas County and the city of St. Petersburg were some of the first governments in Florida to blow out their natural water supply, overpumping groundwater to the point that the sea rushed in where the freshwater used to be. No worries; the governments simply acquired land for well fields to the north and to the east, in largely uninhabited northwest Hillsborough and central Pasco counties. That worked for awhile. But then those counties, too, began to grow. As Pinellas increased pumping to meet its demand, unwitting neighbors who had settled in the rural coun-

A mobile home and shed are swallowed by a sinkhole in Odessa, Florida, in November 1987. Groundwater pumping to quench densely populated areas to the south caused sinkholes, dried-up lakes, and other consequences for rural neighbors to the north. According to the U.S. Geological Survey, more than 80 percent of land subsidence in the United States is a "consequence of our exploitation of underground water." (Courtesy of *The Tampa Tribune.*)

tryside near the well fields complained of dropping lake levels, sinkholes, and dried-up wells. Buddy Blain, longtime general counsel at the Southwest Florida Water Management District, remembered driving out to the well fields to see what was going on for himself. He was taken aback by land subsidence so severe that trees were being swallowed: "As you drive into one place all of a sudden you see the tops of trees that are at eye level," he remembered. "As you got close to it you see that the ground had actually dropped away and made cracks and holes in the land surface."[28]

The political, legal, and even personal battles fought over Tampa Bay's water would fill a book of their own, and they have. Honey Rand, spokeswoman for the Southwest Florida Water Management District during some of the most contentious of the water wars, wrote about them in a book by that name. Rand tells of how, from the early 1970s all the way into the mid-1990s, local governments and water agencies either ignored the families with wells run dry or flat out refuted their claims. Sinkholes, dropping lake levels, and disappearing wetlands were just a natural part of the hydrologic cycle, officials told flummoxed residents again and again. Groundwater pumping had nothing to do with it.[29]

Since this was, after all, a war, it is somehow appropriate that the citizen who finally got through to the water managers was a retired U.S. Army colonel with the Green Berets. Steve Monsees had discovered the bucolic rural lakes of Pasco County while stationed at MacDill Air Force Base in Tampa. In 1988, he bought seven acres of land overlooking a 100-acre lake. There was a pond out back and room for horses. Stationed in Africa during the end of his career, he would make it back to Pasco as often as he could to look at his property. In 1989, he began to build his dream home. When he moved in two years later, the hundred-acre lake was nearly gone.[30]

Various agencies and local governments told Monsees it was just the natural rainfall cycle—that the water would eventually return. But the water-management district's own data showed rainfall had been higher than average.

Monsees blamed his problems on the Cross Bar Ranch Well Field in Pasco County, drilled by Pinellas County government to pipe groundwater south for public supply. Armed with photos of his parched landscape, of dead, upside-down turtles in what used to be the lake, Monsees made a passionate speech to the water management district board in 1994. "I am not complaining to you today of lowered lake water levels, but of the total and complete destruction of all water resources in our community," he said of his neighborhood near the well field. "There is not a parallel in the recorded history of this area, under any drought condition, that approaches the totality of this destruction. All surface water is gone. All wetlands and marshes are gone. Most wildlife has disappeared. The fish and the alligators are gone and now even the trees are dying. No man or woman or government has this right. . . . There has not been a day during the past two years that I have not felt violated and angry because of the loss of our lakes, ponds, wetlands and wildlife. Please restore what is lawfully and rightfully ours."[31]

Plenty of citizens in years before, and, more recently, district staff, had been saying the same thing. But there was something particularly effective about Monsees' plea, maybe about him as a person. His heartfelt speech changed hearts, at least those of the political appointees charged with managing the region's water resources. The Southwest Florida Water Management District's board members finally were convinced that groundwater pumping could not last. The locales surrounding Tampa Bay were going to have to come up with new ways to get freshwater.

Pinellas County and St. Petersburg took a different view. They continued to insist the groundwater was plentiful. They fought the water-management district's threat to restrict their Pasco withdrawal permits, arguing the science was arbitrary and that other users such as agriculture and industry were not being restricted.

For several years more, local governments and water agencies spent millions of taxpayer dollars to duke it out in court. At one point, Pinellas County even sued citizen activists whose wells had gone dry in Pasco. The governments and agencies spent hundreds of thousands more on dueling public relations campaigns—again using the taxpayers' money to sway the taxpayers' opinions. After Pinellas County killed a plan to rehydrate Pasco well fields, a couple of commissioners from Hillsborough and Pasco counties staged a press conference on a boat on the dry bed of Big Fish Lake. The 300-acre lake once produced a statewide record largemouth bass. It had been bone-dry for six years. Hillsborough County Commissioner Ed Turanchik, also chairman of the area's West Coast Regional Water Authority, deadpanned to reporters as he cast his fishing line into dusty weeds: "The plan is to cast for bass, and if that doesn't work, we'll dig for trout."

The litigation alone cost taxpayers more than $10 million over the years, "with not one new drop of water served to the public," Rand observed.[32] The state legislature began to harangue Tampa Bay officials for their abject inability to either end the wars or supply water to their constituents. Senators ultimately threatened that if local leaders could not fix their own problems, state lawmakers would do it for them. That was an unpleasant prospect. Pinellas County Commissioner Steve Seibert, a Birkenstock-wearing Republican who would later serve as Governor Bush's secretary of community affairs, finally held out an olive branch. A lawyer himself, Seibert could see that the only people benefiting from the water wars were the ones billing by the hour to litigate them. He convinced fellow commissioners it was time to concede. "It wasn't going to be pretty," he said. "Someone was going to figure out that it was public dollars fighting public dollars fighting public dollars."[33]

In 1998, six local governments around Tampa Bay declared a truce and established a regional water utility called Tampa Bay Water. The special district sells water wholesale to utilities in Hillsborough, Pinellas, and Pasco counties that in turn sell water to 2.5 million people. Its biggest water-supply project to date, a $148 million seawater desalination plant at the mouth of Tampa Bay, has been an expensive disaster, the subject of more attention in chapter 11. In 2006, three years after it was supposed to start producing 25 million gallons of water a day, the plant was still not up and running. Yet even without the plant, the communities, by working together instead of warring, had slashed their groundwater pumping by a third.

There is a relevant footnote to the Tampa Bay Water Wars. The name of an up-and-coming Pinellas real estate developer who, in the early 1970s, found the Cross Bar Ranch in Pasco County in the first place and negotiated the deal that allowed Pinellas to pipe water from it. The young man was partial to aviator sunglasses and leather jackets. Lee Arnold's deal at the Cross Bar ensured his future development projects would not be held up for lack of water supply.[34]

You would not think Arnold, as a veteran of the Tampa Bay Water Wars, would have been so surprised by Floridians' outraged response to the Council of 100's water-transfer proposal of 2003. But Florida's real estate dealers, indeed the state's entire economic engine, have a short-term memory for bad news. Such oblivion allows them to sell hurricane-prone beachfront condos and sinkhole-prone ranchettes with equally good cheer.

It is often said about Florida that the southerners live in the northern half of the state, the northerners in the southern half. That is generally true, though changing as newcomers fill in even the most rural reaches of the peninsula, and especially as St. Joe builds its upscale communities on a million acres in the panhandle. When it came to the Florida Council of 100's plan for water supply, northern and southern citizens were united in their opposition. Residents of water-rich North Florida, particularly those in counties along the Suwannee, feared pipes or tankers would someday haul their water south to irrigate new golf courses and subdivision lawns in the state's fast-growing metropolitan areas. "There are people in Williston right now getting shotguns and buckshot," to guard their water, said an elected commissioner named Danny Stevens from rural, sparsely populated Levy County, where the Suwannee River flows into vast grass flats at the Gulf of Mexico.[35]

Shotguns aside, Florida's water does not belong to the people who live near it. Rather, it is held in public trust for the benefit of all Floridians. State law has long allowed transport of water "beyond overlying land, across county boundaries, or outside the watershed from which it is taken," where consistent with the public interest.[36] Recent amendments to the law make regions exhaust all possible local sources before poking their straw in a neighbor's water.

Still, even residents who presumably would benefit from a north-to-south water transfer did not want it. The Florida senate's Natural Resources Committee held hearings around the state to air the Council of 100's ideas. During a hearing at the urban, traffic-choked campus of Florida Atlantic University in Boca Raton in southeast Florida, nearly one hundred residents showed up to fend off what they took to be a plot to squeeze more people onto the tip of the state. "We can talk about water supply, water quality, conservation until the cows come home," said an elementary school teacher named Elisabeth Falcone. "The bottom line is: too many people."[37]

Statewide, half of Florida's 67 counties passed resolutions against the plan. Opposition built to a crescendo at the last Florida senate hearing in a rural North Florida town called Chiefland, where more than 1,000 angry residents crammed a high school auditorium, toting signs that said, "Not one damn drop!" and wearing T-shirts that read, "Our Water Is Not for Sale."

OFF A DUCK'S BACK

At the National Hurricane Center on the campus of Florida International University in Miami, 80 staffers monitor Doppler radar from a fortress built with 10-inch-thick concrete walls laced with 45 miles of steel rods. The building is elevated 5 feet above the floodplain, has roll-down shutters and a tornado shelter, and can withstand a direct hit from a 250-pound projectile flying 60 miles an hour.[38] Governor Bush loved the building. He weathered the hurricane that was the Council of 100's water-supply flap so well, it was as if he had been holed up behind those walls of concrete and steel.

One hallmark of Bush's eight-year tenure as governor of Florida was his ability to deflect unpopular projectiles, even ones he had launched himself. During his first successful gubernatorial campaign in 1998, when prodded for specific ideas on environmental and growth issues, Bush had presented a position paper that said he "opposes transferring water from one region to another."[39] That stance never surfaced during the Council of 100's water brouhaha. Bush's earliest comments on the council's plan praised the group for having "the courage to take a position that provokes debate." He declared Florida's water-management system "tired" and "old" and said without changes it would "cost a fortune to continue to be able to grow."[40]

When it became clear that the public would not stand for a water-transfer scheme and the legislature would not touch it, Bush turned his back on the Council of 100 and its water report. Bush similarly distanced himself from the politically unpopular Destination Florida effort he had created. By the spring of 2003, the Destination Florida group had proposed fifty-seven pages of recommendations for making the state more retiree friendly. The centerpiece was a two-pronged marketing campaign to be paid for with state funds. One half of the campaign would work to attract "amenity-seeking baby boomers" aged fifty and over, paying particular attention to "the markets with the highest potential of return, such as New York City and Chicago."[41] For the other half, Florida's taxpayers would foot a campaign to convince themselves of the wisdom of luring new retirees to the Sunshine State. The internal campaign would convince "current residents and policy makers that the value and contributions of its elder residents more than offset the costs of enhancing the state's services and finding better ways to meet current needs."[42] In other words, the state was not adequately funding services for today's retirees and would need to find a lot more money for future retirees. But it would be worth it.

At the state capitol in Tallahassee, Florida's lawmakers had little enthusiasm for the Destination Florida recommendations. They declined to fund a retiree marketing pitch, much less the in-state campaign.

When it came to luring more wealth to the state, Bush would do much better than retirees. During his last years as governor, he orchestrated some of the biggest deals in Florida's history to lure more economically viable growth, such as convincing the Scripps Research Institute in La Jolla, California, to set up a major biotechnology complex in Palm Beach County.

Bush called the Scripps coup "a seminal moment in our state's history," as future-changing as the visions of Henry Morrison Flagler and Walt Disney. State and local governments were poised to spend $800 million in incentives to lure and help build the institute. The governor said it would create 6,500 direct jobs, and that 40,000 more would come from a biotech cluster that would grow around the Scripps campus at Mecca Farms, an orange grove in western Palm Beach County at the edge of the Everglades. To allow a shiny new high-tech city in a sensitive area that the local land-use plan declared should remain rural, county commissioners, of course, rewrote the plan. The Army Corps, meanwhile, rushed through environmental permitting in recognition, according to one Corps e-mail, "of the economic benefits to the state."[43]

In a rare victory for Florida's environmental defenders, a federal judge stopped construction at the Mecca Farms site. Scripps ended up building its biotech hub in an urban part of the county where its scientists were happier, anyway.

Just like he did better than retirees for luring new population, Bush did better than the Suwannee for finding new water for growth. When he left office in 2006, Florida was steering millions of dollars a year toward new water-supply projects statewide. And Bush's accelerated Everglades plan was moving at breakneck speed on major projects to supply water to people and cities in southeast Florida.

For now, the pressure is off the Suwannee River. But the next drought will come. When it does, Florida's thirsty metropolitan areas will be even bigger than before, and the idea of transferring water will pop, like buckshot, into a new developer's head. In times of scarcity, moving water around is as inevitable as it is political. These facts are known all too well by people in many small, river-blessed towns in the American West. The small cities at the headwaters of the Arkansas River in central Colorado, for example, are under enormous pressure to hand Denver pristine water that locals rely on for whitewater rafting and other staples of the ecotourism economy.[44]

The question for eastern states, including Florida, is whether they will learn from the West, where rivers like the mighty Colorado were overallocated until nature and even some legal users were left with no water during times of drought. Would emerging water compacts among states in the East, or the courts, begin to put nature first when divvying up water? That was yet to be seen. But one element of the western experience had clearly moved east: the water wars.

7

Water Wars

THE CHATTAHOOCHEE RIVER begins as a small spring in North Georgia but gains momentum as it travels south, creating pebbled mountain streams and hidden waterfalls that delight southernmost hikers on the Appalachian Trail. The river winds down the entire length of Georgia, meandering through oak and pine forests set off with red maples and white dogwoods, rushing over rocks it has flattened smooth over thousands of years. The Creek Indians gave the Chattahoochee its name: "river of painted rocks."[1] Georgians just call it "the Hooch."

Considering its size and all that it has to do, the Hooch may be the hardest-working river in America. Every summer, the river carries thousands of tourists in tubes or rafts who "shoot the Hooch" through the faux-Bavarian town of Helen and other North Georgia mountain retreats. For fly fishermen, it serves up rainbow, brook, and brown trout stocked by the state.

The Chattahoochee supports sixteen power-generating plants and has fourteen dams.[2] Fifty miles above Atlanta, the Army Corps built Buford Dam in the 1950s and created a huge, brilliant-blue lake called Sidney Lanier. Along with Lake Powell, created by the Glen Canyon Dam, Lake

Lanier is one of the most visited federal lakes in the country. The Chattahoochee keeps the 38,000-acre lake full and makes its 700 miles of shoreline some of the hottest real estate in Georgia.

From Helen in the north to a dam called West Point, 85 miles south of Atlanta, hundreds of municipalities and industries have permits to discharge pollution into the Chattahoochee. Metro Atlanta's sewage-treatment plants dump 500 million gallons of treated wastewater into the river every day. They regularly spill gobs of raw sewage illegally, as well; Atlanta is now under federal consent decree to complete a massive, $2 billion overhaul of its sewer system to stop the spills.[3]

In southern Georgia and Alabama, the river basin helps irrigate nearly 10,000 acres of corn, cotton, peanut, and other crops each year.[4]

When it gets to the Florida line, the Chattahoochee joins another Georgia river, the Flint, to become the powerful Apalachicola. The Apalachicola weaves down Florida's panhandle, first through bluffs, then thick tupelo and cypress swamps. The journey that began at tiny Chattahoochee Spring ends at Florida's huge Apalachicola Bay. There, 16 billion gallons of freshwater a day mix it up with the Gulf of Mexico to create the last unspoiled bay in Florida.[5]

Apalachicola Bay teems with fish and shellfish that the bays of South and Central Florida have not seen in more than fifty years. Wade along its beaches on some afternoons, and you can feel shrimp thump your legs. The bay's shrimp harvest is 6 million pounds a year. It supplies 90 percent of the oysters slurped down in Florida, 10 percent of those consumed in the United States. This productivity requires just the right mix of salt and freshwater: high salinity in the bay brings in predators from the Gulf, enabling them to stalk young marine creatures in their seagrass nurseries.[6]

All of these responsibilities are grave ones for the Hooch. But it has another important job, too. The Chattahoochee is the smallest river in the country to provide water supply to a metropolitan area—in its case, Atlanta.

The city that General William Tecumseh Sherman burned to the ground during the Civil War reigns today as commercial and cultural center of the South. Atlanta hosts one of the busiest airports in the world and corporations such as Coca-Cola, Home Depot, and UPS. It has dozens of colleges and universities, a thriving black middle class, and the

Oysterman Cletis Anderson deposits oysters onto his skiff in Apalachicola Bay in Florida's panhandle. The bay provides more than 90 percent of the oysters slurped down in Florida. The shellfish require the freshwater that flows into the bay from the Chattahoochee River to the north. But Atlanta wants the Chattahoochee, too. (Courtesy of the State Library and Archives of Florida.)

spiciest international flavors between Miami and New York—with warehouselike global groceries where you can buy anything from a daikon radish to a live eel.

But Atlanta may be best known for its sprawling development that has gobbled up thirteen counties in north-central Georgia. The metropolitan area has grown from nearly 1 million people in 1950 to nearly 5 million today. Over the next twenty years, it is expected to be home to another 2.3 million people. That is like the entire population of metropolitan Denver moving in.[7]

In the last decades of the twentieth century, middle-class residents who worked downtown moved farther and farther into Atlanta's burbs to afford bigger and bigger homes. The city's urban footprint doubled between 1990 and 1997, from 65 miles north-to-south to 110 miles.[8] The bumper-to-bumper commuting, the asphalt and the lack of trees all hiked stress as well as smog. For the past 15 years, Atlanta has averaged 40 "code orange" days a year, when the EPA deems the city's ozone levels unsafe.[9]

The new century brought Atlanta dynamic new leaders like Mayor Shirley Franklin, and new ideas about growth. These leaders wanted Atlanta to grow in, not out. Between 2000 and 2005, more than 5,000 people a year chose the city over the suburbs, and many more were expected to follow.[10] Franklin and others were touting a new beltline around the city and other programs to kibosh the car culture. But the inward turn could not fend off the greatest threat to Atlanta. That was no longer SUVs and smog, suburbs and sprawl. It was disappearing water.

Most eastern cities sit atop massive aquifers that supply groundwater for their residents to drink. Or they are close to huge surface-water supplies, such as lakes, that provide drinking water. But Atlanta is not one of them. Groundwater provides only 2 percent of the region's water.[11] For three-quarters of its 450 million gallon-a-day demand, Atlanta relies on the Hooch.[12]

Most of the time, this arrangement works fine. Just like Florida, Georgia is an extraordinarily wet state, with an average 50 inches of rainfall a year. Without complaint, the Chattahoochee usually manages to keep Lake Lanier full, power the energy plants, irrigate the peanut and the cotton farms, and keep just the right amount of freshwater in Apalachicola Bay, even while absorbing sewage spills and supplying all the water needs of 3 million people a day.

But in the late 1980s, the worst drought up to that time dramatically lowered Lake Lanier, making it clear that the Chattahoochee could not meet its ever-increasing demands in times of drought. In 1988, the federal government declared Apalachicola Bay a federal disaster area when reduced flow devastated the oyster harvest. But the Army Corps, which manages Lake Lanier, saw its responsibility to keep as much water as possible in the lake rather than let it flow down to Florida. In 1989, the Corps came up with a new dam and reservoir plan to hold back more of the Chattahoochee in order to supply metro Atlanta during dry times. A year later, the state of Alabama filed a lawsuit to stop the dam, worried that reduced flow would hamper its ability to grow and develop. Florida joined the suit, arguing the upstream withdrawals would harm Apalachicola Bay and the region's signature seafood industry.

These were the first shots in a water war that has no end in sight. It was not unusual that American states were warring over water. But it was unusual that the battle was set in the water-rich East. "The idea that we're having water wars in a region that gets so much rain is astonishing," Aaron T. Wolf, a professor of geoscience at Oregon State University and an expert on water conflicts, told the *New York Times.* "But it is definitely the shape of things to come."[13]

The last time an American state took up arms against another it was over slavery (states' rights, if you insist) during the Civil War, right? Wrong. It was over water—in 1934. Arizona Governor Benjamin Moeur dispatched the National Guard to the Colorado River during construction of the Parker Dam to stop California's "theft" of his state's water. With machine guns mounted on their trucks, one hundred soldiers showed up and stopped the construction. The U.S. Supreme Court later ruled in favor of Moeur's claim that California had acted illegally in diverting water without Arizona's consent. But then, Congress passed a law that made the whole thing legal, anyway.[14]

Water wars are an old story in the American West, about which Mark Twain is oft quoted as saying, "whiskey's for drinkin' and water's for fightin'." (Numerous academics who have searched for the original source of this quote have never found it; thus, it remains in the dreaded "attributed" category, according to Twain scholar Barbara Schmidt.)[15]

When Major John Wesley Powell described, in 1876, the longitudinal line along the 100th meridian that divided an arid West from a moist East, he foresaw "extensive and comprehensive plans" for the West in which "all the waters of all the arid lands will eventually be taken from their natural channels."[16]

The dramatic prediction has come true in some parts of the West, where diversions have completely dried up many rivers.[17] Some years, the mighty Rio Grande, which historically sent powerful surges of freshwater into the Gulf of Mexico, dries to dust before it ever reaches the sea.

To the east of the 100th meridian lay the verdant half of the United States. Powell saw the region so blessed by rainfall and water resources that farmers would never have to irrigate their crops. It would have been unthinkable that inhabitants would ever have to fight over water.

Powell lived in a sparsely populated nation, before, moreover, invention of the electric water pump. He could not have imagined the United States at the turn of the twenty-first century, with nearly 300 million inhabitants pumping groundwater at a rate of 83 billion gallons a day. Even though Americans had built enough reservoirs to hold 450 million acre-feet of water, they had nearly fully appropriated the country's rivers, lakes, and streams.[18]

Water scarcity has now spread from west to east. And with it has come inevitable conflict. Eastern rivers being fought over in recent years included not only the Apalachicola-Chattahoochee-Flint basin but the Potomac, the Savannah, the Yadkin-Pee Dee, the Roanoke, and others.[19] In the nation's heartland, seven states are fighting over the Missouri River in a battle that pits the economic interests of states upriver against those downstream.[20]

"It has been said that 'water litigation is a weed that flowers in the arid West,'" said J. B. Ruhl, a law professor at Florida State University in Tallahassee. "Well, the seeds have blown east."[21]

Two fuels ignite these conflicts in the water-rich East: intense population growth and unclear rules about who is entitled to water resources.[22] One of three solutions usually solves them: Congress, with its authority over interstate commerce, can approve a division of water. Or states can come up with their own water-sharing agreement and enter into a compact. Finally, states can put their fate in the hands of the U.S. Supreme Court, which has "original jurisdiction" in such disputes.[23] For major western rivers such as the Colorado, warring factions have used all these tools to divvy up water. The combination, which in the case of the Colorado comes from more than a dozen different agreements and court cases, becomes known as the Law of the River.[24]

Ruhl believes his side of the country is lucky that it is only now beginning to engage in the conflicts and compacts that build up the Law of the River.[25] In the West, most of the interstate compacts, federal legislation, and Supreme Court cases that divvied up rivers predated modern environmental laws. The ecological impact has been devastating, without enough water left for the salmon or the steelhead in times of drought. News photos from California's last drought showed huge dead salmon piled and rotting on the banks of the spectacular Klamath River, where more than 20,000 salmon died after the Bureau of Reclamation diverted river water to irrigate fields in Oregon and California.[26]

New agreements and allocations in the East, Ruhl says, could recognize ecological values as well as economic ones by setting minimum stream flows that ensure water for nature as well as for people.

That was the core argument of Florida officials for the seven years they and their northern counterparts tried to negotiate an interstate water compact for the Apalachicola-Chattahoochee-Flint basin, known as the ACF Compact. Approved by Congress in 1997, the ACF Compact was the first interstate water compact in the nation since the major federal environmental laws passed in the 1970s, and the first ever in the southeastern United States. "It presented a major opportunity to manage the ACF basin as a system," said Steven Leitman, a consultant with three decades' experience in conservation and water management in the Apalachicola watershed.[27]

But the opportunity was lost. Enormously complex negotiations over the compact's allocation formula stretched from 1997 to 2003. Congress extended the deadline for the formula fourteen times. On the eve of an expected agreement, Florida officials pulled out, insisting Georgia should make greater sacrifices in both reducing its per capita water use and ensuring minimum flows into Apalachicola Bay during times of drought. Florida was making big demands for a state that has only 3 percent of the population in the ACF basin. (Not to mention a state that uses and wastes copious amounts of water.) Georgia has 90 percent of the population in the basin. Robert Kerr, the retired director of the pollution prevention division in the Georgia Department of Natural Resources who represented Georgia in the negotiations, said the core reason the compact broke down was that Georgia "needed far more than the other two states needed."[28]

But Georgia's demand, basically "that the Apalachicola River bear the full burden of any [water] shortage," also seemed unreasonable. Governor Bush put it this way: "Quit stealing our water."[29] The failure of the states to negotiate on their own may not bode well for the goal of securing water for ecosystems downstream. Now, the Law of the Hooch, and that of the Apalachicola and the Flint, will be decided in courtrooms, where, at this writing, Atlanta's lawyers were winning most of the battles. In the fall of 2005, the U.S. Court of Appeals for the Eleventh Circuit reversed earlier decisions by an Alabama federal district court that had stopped metro Atlanta from sucking more water from the Chattahoochee. If the ruling stands, Atlanta eventually will be able to take up to 537 million gallons of water each day out of the Chattahoochee and Lake Lanier. The case is expected to end up in front of the U.S. Supreme Court.

Florida lost another key ruling in the summer of 2006, when a federal judge denied the state's request for more water from drought-dwindled Lake Lanier and other Georgia reservoirs to save threatened and endangered mussels dying on the shores of the Apalachicola River. U.S. District Judge Karon Bowdre of the Northern District of Alabama agreed with the Corps that the mussels were dying from the drought—not Georgia's thirst. "The court cannot hold the Corps responsible for lack of rain," Bowdre wrote.[30]

Water is intensely personal for most people—maybe because we all began life floating cozily in our amniotic sacs. It seems especially personal for people defined by water where they live: those Americans who make a life along the Suwannee and up and down the Mississippi, residents of the Great Lakes states, the watermen who tong their oysters from the Apalachicola and the Chesapeake bays.

Maryland shares parts of the Chesapeake Bay with five other states and the District of Columbia. But Marylanders, much like Floridians near the Suwannee or the Apalachicola, have an iconic connection to the bay and its rivers and streams, which dominate the state's vistas from the bustling Port of Baltimore to the state capital in Annapolis. They have come to see themselves as the rightful caretakers of the region's water and its environment, particularly as they look across the Potomac River at the sprawling, fast-paced growth in northern Virginia and Washington, D.C.

King Charles I gave them good reason to think so. In 1632, he granted Maryland the right to the Potomac River "from shore to shore." For nearly four centuries hence, Maryland and Virginia have battled over control of the river, with Maryland requiring permits for Virginia's use of the water. In 1996, Virginia's Fairfax County Water Authority, which supplies drinking water to more than a million customers, asked Maryland's Department of the Environment to approve a new intake system 725 feet from the Virginia shore. It would pump 300 million gallons of water a day. Under pressure from residents to reject the permit, Maryland elected officials, including then-governor Parris Glendening, did so, "accusing Virginia of an attempted water grab to solve a water-quality problem that was created by its own, pro-development policies."[31]

The response was precisely the same as that of Suwannee residents to the development community's suggestion the tannic river could help ease water-supply woes in South Florida. People in North Florida do not like what they see when they look to the paved-over southern half of the state. Why should they pay for South Florida's planning and development mistakes? "This is as close to North vs. South as you're going to get since the Civil War," the then state senate president Jim King, a Jacksonville Republican, said of the Suwannee salvo in 2003.[32]

Just two weeks later, referring to the Potomac, Maryland house delegate Jean Cryor told the *Washington Post* that "Virginia does not understand what this river is."

"It sees it only through one prism," the Montgomery County Republican said. "It's driven by a philosophy which is, 'Development is everything, and preservation is an interesting theory but I'll get to it one of these days.'"

Cryor, with Glendening's support, tried to pass laws to block Virginia's new water withdrawal. Eventually, a Baltimore judge ruled that the Fairfax water authority could build its intake system.[33]

But Virginia had had enough of being told how it should or should not develop. In 2000, the state filed suit against Maryland in the U.S. Supreme Court, going straight to the top to take advantage of the court's original jurisdiction in disputes between states. The basic question was this: who has rights to the Potomac River? In 2003, the Supreme Court ruled that both sides do. The court said Maryland's undisputed ownership of the riverbed did not give it authority to regulate Virginia's withdrawals, as long as they did not interfere with navigation. The decision was based on two historic documents, including a 1785 compact between the two states that gave citizens of both the right to use their shores.[34]

It was fitting that a 220-year-old document could help resolve a dispute over the Potomac—a river flowing through the very heart of American history. The story proves the power and endurance of water compacts. The nation has more than thirty modern water compacts, involving every major western state, often more than once.[35] To date, only three cover the eastern part of the country. As water conflict flows east, hopefully, cooperation will follow. The Delaware River Basin Commission Compact doles out water to Delaware, New Jersey, New York, and Pennsylvania. The Susquehanna River Basin Compact manages withdrawals in Maryland, New York, and Pennsylvania. But the Great Lakes Basin Compact has the most difficult job. It covers not only the federal government and eight states but the Canadian government and the provinces of Ontario and Quebec. The states and provinces negotiated their compact in the 1940s, and Congress finally approved it in 1968.[36] But the document may never be strong enough to fend off Great Lakes residents' greatest fear: that of outsiders coming to haul their water away.

"The Great Lakes fuel our economy, color our character and literally define the shape of our state," Michigan Governor Jennifer Granholm told her state's legislature in 2004, during a years-long effort by the Great Lakes governors and premiers to strengthen their compact to prevent water exports or diversions. Without tougher laws, said Granholm, formerly the state's attorney general, "those who are already eyeing our treasured lakes as the solution to their water shortages will begin arriving with their pumps and hoses to take their bounty home."[37]

If you were standing on the moon looking at the Earth, you could see the familiar outline of the five Great Lakes. Superior, Michigan, Huron, Erie, and Ontario and their connecting channels form the largest fresh surface-water system on the planet. Covering 94,000 square miles and draining more than twice that much land, the Great Lakes basin holds an estimated 6 quadrillion gallons of water. That is a fifth of the world's fresh surface-water supply, and nine-tenths that of the United States. Only the polar ice caps hold more freshwater.[38]

As vast as they look, the lakes are replenished at an annual rate of only 1 percent. And, just like Florida and Georgia, Maryland and Virginia, and other water-blessed parts of the East, the eight Great Lakes states have begun to battle to protect their signature resource. In this case, the war is with the outside world. "The most water-rich part of the entire world is recognizing that it can't take the resource for granted, either," said Noah Hall, an attorney for the National Wildlife Federation in Ann Arbor.[39]

The biggest of the lakes, Superior, has the largest surface area of any freshwater lake on earth. It stretches 350 miles east to west, 160 miles north to south, and it has a shoreline 2,800 miles long. Its basin drains 49,300 square miles, in parts of Michigan, Minnesota, Wisconsin, and the province of Ontario.[40] Twenty years ago, Minnesota flirted with the idea of piping water from Lake Superior to Wyoming coalfields, where a Texas company planned to mix coal in a slurry for transport south. Not long after, the Mississippi River suffered record low levels, making it impossible for barges to journey south. Speculation abounded that the Army Corps of Engineers would build pipes and transport water from Lake Michigan to the Mississippi in order to buoy the barges.[41]

Fierce public outcry over those schemes led the Great Lakes congressional delegation to convince their peers to put strong new language in the Water Resources Development Act (WRDA). Every two years or so, Congress passes a new WRDA bill to fund the nation's federal water projects. To the 1986 WRDA bill, Congress added this line: "No water shall be diverted from any portion of the Great Lakes within the United States, from any tributary within the United States of any of the Great Lakes, for use outside the Great Lakes basin unless such diversion is approved by the Governor of each of the Great Lakes states."[42] The idea was to buy time so that the U.S. and Canadian governments, the eight states, and the two provinces could work together to beef up their water compact—to build a stronger legal fortress around their precious lakes.

Of course, America's WRDA could not prevent a water grab from the north. And soon enough, a grab came. In 1998, news hit U.S. and Canadian media that a Sault Ste. Marie company called the Nova Group had obtained permits from Ontario to sell up to 160 million gallons of water a year from Lake Superior. The company's plan was to ship the freshwater through the St. Lawrence River and ultimately to undisclosed, water-scarce parts of Asia.[43]

This time, the public response came close to hysteria. Canadians have long been sensitive about water. In 1988, the specter of a parched United States gaining unrestricted access to Canada's lakes and rivers almost defeated the Canada-U.S. free trade deal.[44] Canadian officials warned the permit would set a dangerous precedent under the North American Free Trade Agreement (NAFTA) and the General Agreement on Tariffs and Trade (GATT)—that once water became a commodity, it would be impossible to stop its trade. In the United States, members of Congress were infuriated that Ontario took unilateral action without talking to U.S. agencies. Environmentalists in both countries predicted wholesale draining of the lakes. "If it's open season for all the water in the Great Lakes," warned Sarah Miller of the Canadian Environmental Law Association, "wait and see what happens in the States, where there will be desperate water shortages."[45]

The outcry seemed like overkill considering the amount of water Nova Group had proposed removing. And both NAFTA and GATT allow countries to limit trade in order to conserve "exhaustible natural resources."[46] But at issue was the precedent. Intense public pressure stopped the deal. Just days after Nova Group landed its permit, it agreed to give it up if the two federal governments came up with a joint plan to ban water exports from the lakes. Said a Nova Group company spokesman, "What started out to be a simple idea to help Third World Asian countries in need of fresh water and in turn possibly help the economic climate of Northern Ontario has turned into an international incident."[47]

Affordable transport of freshwater across oceans has so far been the purview of dreamers. Plans to chip giant icebergs and float them to water-stressed areas such as California seem to have melted. Tankers and barges do deliver small amounts of freshwater to water-poor regions willing to pay premium prices, such as the Bahamas and Cyprus. Tankers also supply water during short-term droughts and disasters such as earthquakes. But steep costs prevent their regular use as a source of water supply.[48]

Someday, though, it is plausible that freshwater could be transported across the sea as easily as pipes carry it across land. Since 1997, a British company called Aquarius Water Trading and Transportation has towed water from mainland Greece to nearby resort islands in polyurethane bags that can hold as much as 2,200 tons. A Norway company, Nordic Water Supply, has made similar deliveries from Turkey to northern Cyprus. The problem is the bags are so big they cannot be dragged more than sixty miles or so without puncturing. A California inventor named Terry Spragg thinks he's improved the technology with his patented zippers that link smaller water bags like railroad boxcars.[49]

But after a decade of successful demonstrations, Spragg has yet to sell his "Spragg's bags" for commercial use. He hoped to land a new demonstration project in Florida, where he proposes transferring permitted gallons no longer used by pulp mills in the panhandle down to Tampa and Miami to ease water problems. In North Florida, that idea is sure to go over like a lead . . . water bag. "There are three keys to making it work—the technology, the economics and the politics," says Spragg. "The technology we've validated. The economics we've proven; we know this is cheaper than desalination. It's the politics that's killing us."[50]

Indeed, the very thought makes the Americans and Canadians who share the Great Lakes shudder. A billboard that looms over Interstates 94 and 96 in Michigan sums up the sentiment. It shows a Texas cowboy, a Utah skier, a California surfer, and a sombrero-wearing New Mexican with giant straws sucking up the Great Lakes. "BACK OFF SUCKERS," warns the billboard, paid for by a group called Citizens for Michigan's Future.

But what if the suckers are some of the thousands of children under

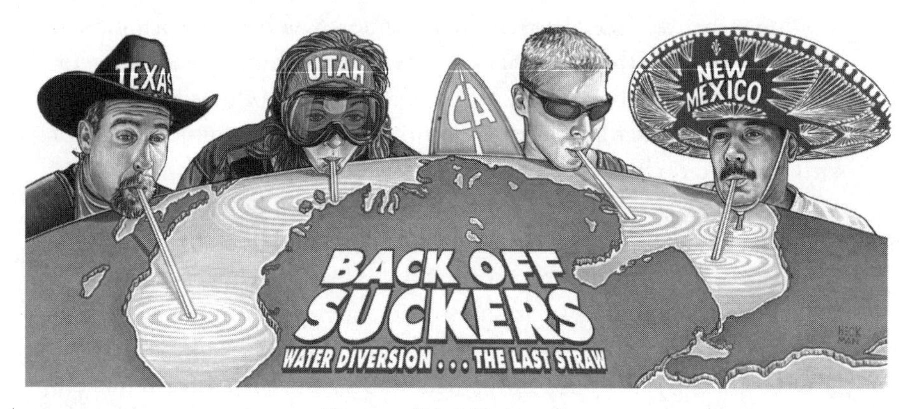

Courtesy of Mark Heckman Art.

the age of five who die every day because they lack access to freshwater? By sea or by pipe, Great Lakes residents do not want to see their water exported to fill a Las Vegas fountain or sprinkle an Arizona lawn. Those types of exports, to be sure, would be worth a tussle. But the global water crisis presents an entirely different specter. If any region of the United States is called upon someday for humanitarian aid in the form of water, it will be the Great Lakes.

Meanwhile, as global demands for freshwater hit the limits of finite supply, experts predict that some day the water wars will become literal. United Nations Secretary-General Kofi Annan has predicted that "fierce competition for fresh water may well become a source of conflict and war in the future."[51] Just since 2000, conflicts over water have led to violence in Ethiopia, Afghanistan, India, China, Israel-Palestine, Pakistan, Macedonia, the Philippines, Nepal, Colombia, Sudan, and Iraq—in some of those places repeatedly. But the Pacific Institute, which tracks these water conflicts, also found that water is far more often a source of international cooperation.[52] Water conflict does not have to lead to war. In the words of Israeli leader Shimon Peres, "If roads lead to civilization, then water leads to peace."[53]

But peace begins at home, and it was not the path regions such as Florida-Georgia-Alabama or the Great Lakes states were choosing at the turn of the twenty-first century. Instead, warring factions were taking all-or-nothing stands. The small Wisconsin community of Waukesha was just one place in danger of ending up with nothing. A former resort dotted with more than seventy lakes, Waukesha was once so well-known for its mineral springs that it earned the nickname Saratoga of the West.[54] Today, it is running out of water. Groundwater pumping has caused Waukesha's aquifer to sink 600 feet. The water that deep is contaminated with cancer-causing radium. Waukesha County sits five miles west of the Great Lakes basin boundary. Some residents can actually see the shoreline of Lake Michigan from their streets. Yet Great Lakes neighbors are fighting Waukesha's proposed diversion of Lake Michigan water to solve its crisis. Under the federal ban on diversions, Waukesha needs approval from all eight Great Lakes governors. Environmentalists and other opponents argue that the problem, again, is the precedent. "If we say yes to Waukesha County, it's hypocritical to say 'no' to the West, or Asia," said Cameron Davis, executive director of the Lake Michigan Federation.[55]

Clearly, the Great Lakes Basin Compact needed an update to allow some neighborliness while tightening restrictions against profiteering water exporters. In the wake of the Nova Group's proposal, the Great Lakes governors and the premiers of Ontario and Quebec met in Niagara Falls, New York, in 2001 to begin work on a new compact before the next water hauler comes knocking. The effort, known as Annex 2001, led to a historic accord on diversions that the leaders signed in December 2005. The agreement bans big water diversions outside the Great Lakes basin. But it opens the door slightly for local communities such as Waukesha that are suffering water shortages in plain sight of the greatest source of freshwater on the planet.[56]

The updated compact now must be approved by all eight state legislatures as well as Congress. It is not at all clear how members of Congress from the West and the Sunbelt will view it.

There is a little matter of hypocrisy, since the city of Chicago and its suburbs divert nearly 2 billion gallons a day from Lake Michigan under a 1967 Supreme Court decree.[57] And then another type of exporter is pumping away, too. Despite the extraordinary, international effort to ensure profiteers can never take water out of the basin, one enterprising industry has a booming business exporting water from the Great Lakes. Its companies pump up Great Lakes water, transport it to other parts of the nation, and sell it for big profits. They do not haul the water in barges or float it in giant bags or send it through huge concrete pipes. They pack it into very small bottles, just right for tucking into your purse or balancing in your car cup holder.

8

Business in a Bottle

TUCKED IN A hardwood forest in rural Madison County near the Florida-Georgia line, a spring called Madison Blue bubbles up into a limestone basin along the Withlacoochee River. Popular with divers and local kids who leap off its wooden ledges and shoot down its short run, the spring pool is only 40 feet wide and 25 feet deep. But each minute, it pumps 45,000 gallons of cold, clear springwater. Poets call this water liquid light. To the bottled-water industry, it is liquid gold.

Madison Blue is one of 33 "first-magnitude" springs in Florida, a designation for the very largest springs in the world, those that discharge at least 100 cubic feet a second, or some 65 million gallons a day, about double the daily water supply for a medium-sized American city. Florida has more first-magnitude springs than anywhere else on the planet. That fact, along with Madison County's proximity to major U.S. trucking arteries Interstate 75 and Interstate 10, led a huge, multinational corporation to bring its operations to the tiny county's tiniest town—Lee.

Motto "Little But Proud," Lee is one of the last outposts in Florida where you can drive for miles and miles on graded dirt roads and never see another soul, save for the occasional chicken in the road. But today in these woods, not far from trailers with no trespassing signs like the one

A rite of summer in North Florida, boys leap into Madison Blue Spring, near the Georgia border. Nestlé Waters North America pumps Madison Blue for its Zephyrhills and Deer Park bottled-water products. The company donated money that allowed the state to turn Madison Blue into a park.
(Courtesy of John Moran.)

that says, "BAD ASS DOGS," sits Nestlé Waters North America's newest plant, one of the most state-of-the-art bottling facilities in the United States. The plant cranked out 26 million cases of Zephyrhills, Deer Park, and Nestlé Pure Life products in 2005. It is also the southeastern U.S. distribution center for all of Nestlé's water products, from the French Perrier to the Italian S. Pellegrino.[1]

Nestlé came to town at the crest of a wave of new water-bottling plants throughout the eastern United States. In Florida in the early 1990s, only one small company pumped and bottled water inside the boundaries of the Suwannee River Water Management District, which oversees the famed river and many of North Florida's springs, including Madison Blue. Toward the end of the decade, the district's regulators saw more than a dozen new applications for permits to withdraw springwater to bottle or sell. Most came from land-owning families trying to get in on what looked to be a lucrative fad.[2]

Bottled water would be more than that. Whether seeking taste, convenience, or a healthier choice than soda, or (often wrongly) assuming it was better for them than tap water, Americans by 2000 had started forking out more than $6 billion a year for bottled water. That year, the average American drank 17 gallons of bottled water. Five years later, it was 26 gallons.[3]

In 2003, bottled water became the second-highest-volume commercial beverage in the United States, behind soft drinks. Its enormous U.S. growth is often linked to that year's outbreak of cryptosporidium in Wisconsin tap water, which made 400,000 people sick and garnered sensational headlines across the country. By 2005, U.S. sales of bottled water had hit $10 billion.[4]

With an eye to the steady 10 percent annual increases in sales, bigger and bigger companies came looking to tap the Suwannee region's springs. Land and permits began changing hands from families to large corporations. Today, three of the largest water-bottling companies in the world pump or buy their product from the Suwannee district: Nestlé, a unit of Swiss giant Nestlé SA, the world's biggest food and beverage company; Atlanta-based CCDA Waters, owned jointly by Coca-Cola and French food and beverage powerhouse Danone; and DS Waters of America, the top company for home and office water delivery in the United States.

The attractions for bottlers to Florida are as numerous as for tourists to Orlando. For one, the companies avoid long transportation hauls by locating operations close to a large, thirsty consumer base. Analysts say high shipping costs mean bottlers want to be as close as possible to population centers. "Florida has become one of the top-consuming states for bottled-water in the United States," says Gary Hemphill, managing director of the Beverage Marketing Corporation in New York. "It is associated with good weather, outdoor activities and an active lifestyle."

Bottling companies also get little oversight in Florida. The Division of Food Safety, the state agency that monitors the water companies, does some testing of bottled water to make sure it is safe and inspects bottling facilities for sanitation. State law also requires the division, part of the state Department of Agriculture and Consumer Services, to ensure that bottled water is "from an approved source." But the food-safety regulators say that simply means they check that the companies have approval from an agency such as a water-management district to withdraw water.

No state agency monitors precisely what the companies are bottling. No one determines whether the companies are bottling "springwater," as opposed to the groundwater that 92 percent of Floridians get out of their taps.

At the national level, bottled water and tap water are regulated by different federal agencies. The EPA regulates municipal water under the federal Safe Drinking Water Act. The Food and Drug Administration regulates bottled water under the Federal Food, Drug, and Cosmetic Act. FDA rules cover the source, safety, and labeling of bottled water. The agency is supposed to inspect bottled-water facilities, but generally they get low priority because of their relatively good safety record compared to food plants. The bottlers also do their own sampling and testing. But many do so far less frequently than municipal suppliers.[5] While Nestlé officials say they test water about 100 times a day, for example, the director of quality and technical services for DS Waters says his company tests 4 times a year.[6] By comparison, New York City tap water was tested 430,600 times in 2004 alone.[7] Regardless, FDA officials say they do not concern themselves with water that never crosses state lines. That goes for much of the product bottled in Florida.

Besides its bubbling springs, there is something about Florida that bottlers do not find in lots of other states where they operate: its bubbling politicians. Many love the industry, with its promise of jobs and multimillion-dollar plants. To be sure, the bottlers face some NIMBY (not-in-my-backyard) activism in Florida. But it is a drop compared to that seen in other parts of the United States and Canada. In Florida, handing over resources in exchange for economic development has been a part of the state's heritage since the legislature traded all that swampland for railroad lines in the 1800s. Elected officials, especially those in small, rural counties such as Madison, where 23 percent of residents live below the federal poverty line, want jobs more than they worry over environmental consequences, regardless of Florida's water-supply problems.[8]

They are as warm and welcoming as a postcard from the Sunshine State: Come on down. The water's . . . free.

It is a nice reception compared to the cold Midwest.

Home to a healthy population of wild brown trout—sleek, bronze game fish spotted black, blue, and red—the Mecan River in east-central Wisconsin is one of the top fishing destinations in the state. But to water bottler Perrier Group, the real catch was a group of large, clear springs at the river's headwaters.

In 1999, Perrier and Wisconsin officials met to discuss the company's plan to sink a well at Mecan Springs, a state-owned conservation area managed by the Wisconsin Department of Natural Resources. Company officials wanted to build a million-square-foot bottling plant in the rural town of Richford nearby. They promised 250 well-paying jobs.

Town residents, environmentalists, and the state's powerful sportsmen's lobby were so ardent in their opposition that even Republican Governor Tommy G. Thompson, a Perrier ally, did not dare help the company wrest permits for Mecan Springs. And so Perrier turned, with its promise of jobs and doubling the tax base, to another small, rural community: New Haven, Wisconsin, home to Big Spring.

Perrier made sure there were no beloved trout streams, and therefore no wealthy fly fishermen, to contend with this time. But the company underestimated the people of New Haven. If Perrier officials could have bottled up the virulence, they could have sold it in cases as weapons to the Department of Defense. Some New Haveners were so angry about the company's plans that the county sheriff, Roberta E. Sindelar, feared violence against the local farmer negotiating to sell land to Perrier for its plant. "The terrible thing is, I think something is going to happen," Sheriff Sindelar told a *Milwaukee Journal-Sentinel* reporter. "It's an explosive issue."[9] In the end, then-governor Thompson urged Perrier to abandon its Badger State plans.

It is not easy, industry officials say, to find a reliable, good-tasting source of "springwater" that meets the FDA's definitions for bottled water labeled as such. At the turn of the twenty-first century, Nestlé Waters North America, parent company to Perrier Group of America, dispatched natural resources managers all over the eastern United States to scout springs for new operations close to population centers. The company particularly needed bottling plants and distribution hubs in the Midwest and the Southeast; at the time, its Allentown, Pennsylvania, facilities served both regions.[10]

For its southeastern plant, Nestlé would settle on Madison County in Florida. There, local officials courted the company, and state leaders helped pave the way with a $1.3 million transportation grant to build a connector between the plant and the nearest highway. For a midwestern presence, after its rude boot from Wisconsin, the company turned its attention to two springs in Michigan, north of Grand Rapids. Residents of the area, which is known as the Tri-Lakes, immediately organized. They formed Michigan Citizens for Water Conservation. Their ranks soon grew to more than 1,000.

In August 2001, the Michigan Department of Environmental Quality granted Perrier Group permission to pump up to 400 gallons a minute from a source called Sanctuary Springs, which feeds the Little Muskegon River, a tributary to Lake Michigan. The following month, Michigan Citizens for Water Conservation filed a lawsuit, arguing, among other things, that the company's pumping of the springs would have an adverse impact on nearby surface waters.

At the same time, Jennifer Granholm, then Michigan's attorney general, was looking into whether Perrier's plans violated the U.S. Water Resources Development Act's ban on diversions. The ban covered not only the Great Lakes themselves but "all streams, rivers, lakes, connecting channels and other bodies of water, including tributary groundwater, within the Great Lakes basin."[11]

Nestlé went forward with its state-approved plans, and on May 23, 2002, opened its $150 million plant and began bottling Ice Mountain brand springwater, pumped from Sanctuary Springs. Over the following year, it continued to battle Michigan Citizens for Water Conservation in the state's 49th judicial circuit, in front of a Mecosta County circuit judge named Lawrence C. Root. Judge Root was a native of the county, hailed from a manufacturing family, and had an undergraduate degree in business administration. No one expected his astonishing, 67-page decision of November 2003. Calling the case "the most extensive and intensive" in the history of the circuit, he wrote that "Nestlé's pumping operations at the Sanctuary Springs *must stop entirely.* I realize this is a dramatic and drastic result, but from the evidence I accept . . . I am unable to find that a specific pumping rate lower than 400 gpm, or any rate to date, will reduce the effects and impacts to a level that is not harmful."[12]

Michigan's bad news for Nestlé did not end there. In the fall of 2002, voters had sent Granholm to the governor's mansion. As Great Lakes residents and politicians became increasingly paranoid that bottling would open the door for other businesses that wanted to ship water out of the region, Granholm came under great pressure to strengthen the state's diversion laws. In an executive order, she banned Nestlé from selling its Michigan-bottled products outside the Great Lakes. Under the U.S. Water Resources Development Act, she argued, the governor of any Great Lakes state could veto water diversions outside the basin. She also slapped a moratorium on new or expanded bottled-water operations in the state until the legislature could enact a water-withdrawal law.[13]

They were surprising actions from a governor who won office touting economic recovery, who would base her reelection campaign around a plan called "Jobs Today, Jobs Tomorrow."[14] Michigan in 2006 suffered its sixth straight year of job losses, a string unprecedented since World War II. From 2000 to 2005, the state lost more than 300,000 jobs as its manufacturing workforce was slashed by more than a quarter.[15]

But, as Judge Root had felt compelled to say in his ruling: "Michigan is a state in which tourism is a major part of the economy and many people who choose to live here do so because of the recreational opportunities in and natural beauty of the state, much of which has to do with our aquatic resources, of which many here feel very possessive."[16]

The same words could have been written about Florida, a place that is gaining rather than losing jobs. Fueled by population growth, Florida's jobs grew by more than 3 percent each year from 2000 through 2004.[17] That is more than twice the rate of U.S. job growth in the same period. Some people argue that Floridians simply do not feel the state pride or ownership felt by residents from the Great Lakes State, or the notoriously boastful Lone Star State of Texas, because so many of them migrated to Florida. Eighty percent of residents were born somewhere else.

But that is not a fair assessment of a population whose tax dollars fund, as just one example of stewardship, the most expensive conservation-land-buying program in the world. Like overindulgent parents who hand their son the keys to a Porsche on his sixteenth birthday, the state's politicians have convinced residents that Florida can have it all: fat tax coffers from the dizzying population growth, a constant stream of jobs from new businesses, and natural beauty that, while shrinking, is still stunning. Especially if you just moved from, say, Gary, Indiana. Watching a fiery sunset from any beach along the Gulf of Mexico remains a spectacular experience. As long as you keep your back to the sky-blocking condos that likely loom behind you.

In Xanadu did Kubla Khan
A stately pleasure-dome decree:
Where Alph, the sacred river, ran
Through caverns measureless to man
Down to a sunless sea

So begins one of the most famous poems in the English language, Samuel Taylor Coleridge's "Kubla Khan." Coleridge admitted he composed the poem in a profound opium sleep and never visited any of the places he evoked so powerfully. He was drawn to them by a diverse collection of other writers, one being William Bartram, the American naturalist who traveled extensively in Florida in the late eighteenth century and wrote so ebulliently of the state's springs. Coleridge scholars have traced his "caverns measureless to man" directly to the springs of North Florida as described in Bartram's *Travels.*[18]

Wrote Bartram in his *Travels,* published in 1791: "We now ascended the chrystal stream; the current swift: we entered the grand fountain, the expansive circular bason, the source of which arises from under the bases of the high woodland hills, near half encircling it; the ebullition is astonishing, and continual, though its greatest force or fury intermits, regularly, for the space of thirty seconds of time."[19]

Wrote Coleridge in "Kubla Khan":

And from this chasm, with ceaseless turmoil seething,
As if this earth in fast thick pants were breathing,
A mighty fountain momently was forced:
Amid whose swift half-intermitted burst
Huge fragments vaulted like rebounding hail[20]

Two hundred years after Bartram described them, Florida's springs can no longer be said to bubble with "astonishing ebullition," or even force. To be sure, the springs are still inspirational to writers and other artists who try to capture their magic in words or paintings or photographs. But these days, they are as likely to draw industrial engineers as poets.

Back in the woods of Madison County, outside a sprawling plant painted three shades of blue, high-tech water bottling begins in three shining silos. Each holds 60,000 gallons of water. Only some is springwater. Piped from Madison Blue a mile away or trucked in from other parts of Florida, the springwater will end up in Zephyrhills or Deer Park bottles. The rest is simply groundwater sucked from underneath the bottling plant. It will become Nestlé Pure Life water—plain or fruit flavored.

Inside, the spotlessly clean plant is a 650,000-square-foot behemoth of stainless steel. Bottles bob in bunches along conveyor belts that carry them up, down, and around in a computerized ballroom dance that repeats twenty-four hours a day, seven days a week. Only a few workers can be seen, hair covered with nets, ears plugged against the noise. Several are tasting and testing in a glass-walled laboratory.

Pumped from a silo into the plant, the water rushes first through a series of microfilters to remove particles. It is sent through UV light for sterilization. A low dose of ozone kills microorganisms. Then it heads to one of four manufacturing lines that whir and clack. Two of the lines run at 1,200 bottles a minute; the two others at 510 bottles a minute.

The bottles themselves start out as three-inch plastic tubes called "preforms." By the thousands, the preforms are dumped from cardboard boxes into a small elevator that hikes them up and drops them into a machine that turns them right-side up. Then they roll along the production lines into a "blow-molder," which warms them with light and blasts them with heated, sanitized air that stretches their plastic molecules into a perfect, half-liter bottle.

The freshly made bottles rush through a machine that fills them with the filtered water. They bob into another that twists on their caps, another that makes sure the caps are secure and fill heights accurate. Next, the conveyor carries them to a "canmatic" that glues their labels perfectly as they spin. A laser etches a code on each bottle. They travel along a wet belt through a lubricant that keeps them from tipping over. Hundreds now move through one of four, narrow openings that send them into a vast packaging and storage area, where forklift drivers with laptops and scanners move cases from tall stacks to tractor-trailers.

From underneath conveyor belts on the packaging side, corrugated cardboard swoops up to encase the bottles; an automated glue gun pops up to fasten the sides. Another machine slaps on film wrap, and the cases move through a heat tunnel for the packaging to set. "Deer Park," they say on this day. "Taste the spring water difference."

So just how different is springwater? To the ancients, natural springs were thought divine. Greek and Roman doctors wrote about their healing powers. For thousands of years, people throughout the world turned to "water cures," submerging in springs for therapy.[21] Well into the twentieth century, wealthy northerners would come to Florida to soak in its pane-clear springs, believing the 70-degree water would keep them young and vital.

But today, Florida's springs are not so healthy. In recent years, almost every spring in the state has shown degraded water quality. Scientists refer to a once-spectacular spring called Volusia Blue as "Volusia Green" because algae have turned it to pea soup. The problem has shown up all over Florida as the state's storm water, farms, spray fields, and septic tanks carry nitrogen and other pollution into springs.[22] Scientists from the U.S. Geological Survey recently found low levels of DEET in Florida's springs.[23] "There's nothing magical about spring water," says Angela Chelette, chief of groundwater regulation at the Northwest Florida Water Management District in Florida's panhandle, home to about a half dozen of Florida's springwater bottlers. "There's some pretty nasty stuff in there."

For the most part, Florida businesspeople have given up on hawking springwater as curative—a bottled adaptation of the Fountain of Youth myth. However one company, Golden Springs LLC, which runs a spa at Warm Mineral Springs in North Port, pumps and sells "Fountain of Youth Mineral Water" for $9.95 a liter. The company's promotional director, Robin Sanvicente, says the bottles sell well to people all over the world. "It rejuvenates, replenishes, restores, actually heals, arthritis, fibromyalgia, you name it," she says.

Few other bottlers in Florida make such outrageous claims. But even figuring out whether you are buying "pure" Florida springwater is impossible. The state's environmental scientists and regulators do not agree where groundwater stops and "springwater" begins. Some argue it is not springwater unless an intake pipe pokes into the spring itself. Others say as long as the well is in the spring's "zone of influence," the water is identical.

In several cases, the wells of springwater companies are thousands of feet from the actual spring. Nestlé's borehole for Madison Blue is nearly 5,000 feet from the spring. Federal regulations say, "spring water shall be collected only at the spring or through a borehole in the underground formation feeding the spring." The companies must hire a licensed hydrogeologist to show the FDA the groundwater they are pulling up is of the same composition and quality as that flowing from the nearby spring.

To Chelette, the groundwater expert, the water in her district is the same, whether consumers get it from their faucets or pay a dollar a pint to drink it from a bottle. "I would call it groundwater. It's the water that we drink out of our taps and our spigots," she says of the source water for bottlers under her purview. "They may run it through a couple of more processes, but generally, it's all the same. It's good water—we all have good water."

Of course, much of the water bottled in the United States including Florida makes no claim to being springwater. Called variously "drinking water," "purified water," or "natural water," many products are just municipal water poured from the tap or groundwater pumped from wells. An estimated quarter of all bottled water in the United States begins life as tap water. The top-selling bottled water in the United States, Pepsi's Aquafina, is simply tap water that has been additionally purified via reverse osmosis and carbon filtering. The same goes for the number two product, Coca-Cola's Dasani, which counts Jacksonville tap water among its sources.

Labels and marketing, of course, often suggest a source more exotic than the water's origins. Sometimes, the gap between what's on the label and what's inside can be so obvious as to be humorous. Aquafina, whose many municipal sources include the Detroit River, features snow-capped mountain peaks on its labels. Everest Water is not from Mount Everest but from Corpus Christi, Texas. Glacier Clear Water is tap water from Greeneville, Tennessee.[24]

The labels on Crystal Springs Natural Spring Water, bottled and marketed by Atlanta-based DS Waters of America, feature snow-covered mountain ranges, too. The source for the bottles sold in Florida is Wekiva Springs, in the tiny, decidedly nonmountainous north-central Florida town of Gulf Hammock.

Then there is Silver Springs Bottled Water, which calls itself Florida's largest privately held bottled-water company. The firm's name harkens the deep blue springwaters of North Florida and the longtime tourist attraction that is the largest artesian spring formation in the world. But the company uses Ocala well water for many of its products, according to its water-use permit from the St. Johns River Water Management District.

Industry officials say the added value in bottled water, regardless of its source, comes from purification processes, the lack of chlorine, or the fact that their water does not travel through old, municipal pipes. "Bottled water provides consistent safety, quality and good taste," says Stephen Kay, vice president for communications at the International Bottled Water Association. "Consumers like that consistency."

Taste, as the Deer Park label suggests, may indeed be the biggest difference between bottled and tap water. Dissatisfaction with the taste of locally available tap water is the most common reason offered to explain the growing consumption of bottled water.[25] At an Atlanta event famous for its tongue-in-cheek rating of municipal waters in the Southeast, ratings ranged from 0 (sludge) to 13 (nirvana). Memphis won with comments such as "on the nose, at first it was cottony . . . a refreshing texture." Atlanta's water was like "a gulp of a swimming pool." Judges said Houston's tasted "like a chemistry lab." Charlotte, North Carolina, water was said to taste "like a wet Band Aid." Of Orlando's, the judges said, "It's the reason most people don't drink water."[26]

Notoriously, plenty of taste tests have found consumers unable to distinguish between bottled and tap waters. But perhaps more importantly, tests show that bottled water is no healthier than tap water. In fact, some of it contains just the sorts of disinfection byproducts that people buy bottled water to try to avoid.

In 2006, *Florida Trend* magazine sent a half-dozen water samples to Ohio-based National Testing Laboratories, the largest provider of analytical services to the U.S. bottled-water industry. A drinking-water-quality expert, University of Florida environmental engineering professor David Mazyck, interpreted the results. He found that Florida's tap water samples were just as good for you as the bottled-water samples. He also found that bottled waters are not all the same.

Florida Trend tested Orlando tap water, which comes from groundwater in the Floridan Aquifer, and West Palm Beach tap water, whose origin is Lake Okeechobee. The difference between those and most of the bottled products was the presence of trihalomethanes, or THMs, a common byproduct of drinking-water disinfection linked to increased risk of cancer. In both Orlando and West Palm Beach tap water, the THM levels were small. They were 0.020 milligrams per liter, while the

EPA's maximum level allowable in drinking water is 0.080 milligrams per liter.

What might surprise consumers who buy bottled "drinking water" is the presence of THMs in some bottled products, too. *Florida Trend's* test of Publix grocery store brand drinking water found precisely the same level of THMs (0.020 milligrams per liter, again, considered safe by the EPA) as was in the tap waters tested. Publix spokeswoman Maria Brous says the bottle *Florida Trend* tested began life as Atlanta tap water.

Three bottled springwaters were tested by *Florida Trend*, and none revealed traces of THMs. But that does not mean they were free of disinfection byproducts. In the test result most troubling to Professor Mazyck, a sample of Crystal Springs Natural Spring Water, the DS Waters of America product pumped from Wekiva Springs, contained the EPA's maximum-allowable level of bromate, another disinfection byproduct linked to increased risk of cancer. The Crystal Springs bromate level was 0.010 milligrams per liter. That is the highest level of the contaminant that the EPA would allow in drinking water. Kent Kise, director of quality and technical services for DS Waters, says the level detected did not worry him because the federal standards are rigorous to ensure no risk to consumers. "It meets all regulatory standards," Kise says, "this is why we have standards."

Most consumers have probably never heard about the issue of bromate in springwater, but Kise says it is "of very high interest to the bottled-water industry as a whole." Bromate does not occur naturally in springs. Its harmless cousin, bromide ion, can occur in springs, sometimes as a result of saltwater intrusion. Bottlers use a purification process called "ozonation" to ensure water is free of bacteria. When bromide is present, the ozonation process can turn the harmless ion into carcinogenic bromate. "I don't think consumers realize that bottled water can have disinfection byproducts," says Mazyck, "and that that can be the case even if the bottle says 'spring water.'"

The water that passed *Florida Trend's* test with the highest marks was Nestlé's Deer Park Spring Water. But overall, says Mazyck, outside of the bromate issue, "if you drank two liters of water from any of these sources every day for your lifetime, you wouldn't have any adverse health effects."

"That goes for the municipal water and the bottled water. You can't conclude that one is healthier than the other," Mazyck says, although he asserts that EPA oversight of municipal water plants is more stringent than FDA's regulation of bottled water.

Mazyck's overall conclusions are similar to those of other studies comparing bottled and tap waters. In a blind study using ten municipal and bottled-water samples from Central Florida, James Taylor, director of the University of Central Florida's Environmental Systems Engineering Institute, found that both types of water met state and federal water-quality regulations. Two of the bottled waters had high bacterial counts. The municipal waters had significantly higher chlorination byproducts. Overall, says Taylor, there was virtually no difference except that bottled water "costs 10,000 times more."

Does the content of bottled water really matter to consumers? Continued soaring sales indicate Americans care more about things like convenience and calories, style and status. Hemphill, the beverage analyst, says he thinks consumers base their bottled-water decisions on three primary factors: convenience, the packaging, and the price. "Whether it has a sport cap or a twist-off cap," he says, "is often more important to the consumer than whether it's drinking water or springwater."

At the elegant Ristorante Bova in Boca Raton, on Florida's southeast coast, water falls gently down a glass wall as customers in the all-white dining room ponder a two-page water menu. The choices go far beyond "sparkling" or "still." Owner Anthony Bova offers twenty-five bottled waters, from Ty Nant ("for the style-conscious," the menu says, "known for its beautiful royal blue presentation bottle and tiny light bubbles") to Hildon ("English mineral water has a well-balanced, clean, pure taste and has become a byword for style in the restaurant and hotel world"). "It is fashionable," says Bova, whose bottled offerings begin at $6.75 a liter. "The design of the bottle is almost as delicate as the high-end wine bottles or the Italian liqueurs."

James Twitchell, author of *Living it Up: America's Love Affair with Luxury Goods,* calls bottled water a perfect example of the "status marketing" that has helped spending on luxury goods in the United States grow four times faster than overall spending. Like $25 cashmere socks, he says, there's no real reason to buy costly bottled water, except that it feels good to do so. "We're not buying a bottle of water," he says. "We're buying a sensation about ourselves."

Nestlé's investment in rural Madison County now tops $110 million. County Commissioner Roy Ellis, who represents the town of Lee and helped land the plant, says it has been a blessing. Offering incentives and making it easy for companies to do business is the only way Florida's rural counties will ever have the chance to grow and prosper like the rest of the state, he says. "They've been a very good neighbor and they've kept every promise and they've filled every job they said they'd fill."

Ellis points to Nestlé's relatively small daily withdrawal from Madison Blue—1.4 million gallons. He says that is not likely to have an adverse impact on the environment when the spring produces almost one hundred times that much.

Few people realize that what Nestlé takes from Madison Blue is only a fraction of what it pumps and trucks from elsewhere in the state to bottle at the Madison plant. Nestlé also has permits, from a different water-management district, to pump from Cypress Springs and White Springs to the west, in the panhandle. Meanwhile another water-management district, in southwest Florida, permits it to pump Crystal Springs north of Tampa. Add up all its permits, and Nestlé could bottle closer to 5 million gallons a day of Florida's springwater. That does not include the millions more it buys from water dealers in the state, and millions more in groundwater that it pumps from wells to produce drinking-water products.

Some experts cite lower water tables, saltwater intrusion, and a disruption in habitat for fish and other wildlife as environmental problems associated with the bottling industry.[27] But the state's water managers agree with Ellis, pointing out the industry's overall withdrawal rates are miniscule compared to others'. For example, North Florida's pulp and paper mills withdraw an average 155 million gallons of water a day; all the springwater permits combined add up to a little more than 10 million gallons a day. In addition, water managers say, some bottlers help protect springs by buying up surrounding land and keeping out things like subdivisions.

What about the risk of punching boreholes into the karst geological formations where springs are most common? Regulators say there is always a risk of subsurface collapse, but it is usually not enough to stop them from granting a permit. They refused on those grounds only once, at a popular manatee hangout called Three Sisters Springs, because a collapse could have deprived the endangered sea cows their preferential use of the spring's shallow shelf.

The bottom line, regulators admit, is that they simply do not understand the overall environmental impact the water-bottling industry has on Florida. The state is now home to more than sixty bottling-related companies. They range from boutique firms that will slap a company logo on bottles of water to "water dealers" that pump water from various parts of the state and sell it to bigger bottlers. What begins as a free resource passes from dealer to bottler for some 5 cents a gallon. Consumers, of course, eventually pay some one hundred times more.

Why should the $66 billion Nestlé or other bottlers get their raw material for free? One economist compares it to a food company that makes berry jam and gets the berries at no charge. Others believe the government should receive royalties, such as those paid by oil companies, for letting bottlers extract the planet's most important natural resource.

Nestlé and other officials counter that theirs is a value-added product that relies on water to a lesser extent than competitors such as soda or beer. Company officials say it takes 1.3 gallons of water to produce 1 gallon of Nestlé springwater, compared to 3 gallons to make a gallon of soda or 42 gallons to make a gallon of beer. "We pay a great deal for this water," says Meg Andronaco, Nestlé natural resource manager for the southeastern United States. "It costs millions and millions of dollars just to develop the spring and go through the permitting process."

That argument no longer holds water in some other parts of the United States. In Maine, Nestlé pays 0.6 cents a gallon to the state for the water it obtains from Poland Spring. A citizens' initiative underway in that state proposes a 20-cent-a-gallon tax for all large water withdrawals, a move Nestlé officials say will ensure the company's departure. In the Great Lakes region, some states are beginning to assert that "tradable goods," including food products such as beverages and processed foods, are different from pure water, an essential element of life, or "public goods."[28]

It seems like a no-brainer to charge corporations a little something for the groundwater from which they profit. Especially in places like Florida, where other users are being asked to cut groundwater pumping, and where citizens face steep water-rate hikes to cover costly alternative water-supply schemes such as desalination plants. When Nestlé was working its way through the Madison Blue permitting process in the midst of Florida's drought in 2000, a Central Florida citizen named Brad Willis wrote the governor with a relevant question: "I must ask you,

Governor Bush, why your administration has not looked into the issue of foreign/multinational corporations coming into Florida, taking our water for free, and selling it back to us for a huge profit, when honest, hardworking citizens of this state must observe water restrictions and pay ever-increasing water bills," he wrote. "Also, why are these corporations allowed to keep pumping unabated during rationing, a time when the citizens of this state are not even allowed to wash their cars after 10 a.m., or not at all on some days?"[29] The governor said he would look into it.

In 2005, a work group studying ways to fund new water-supply alternatives in Florida did consider lifting a sales-tax exemption on bottles of water, which would have raised an estimated $50 million a year. Around the same time, the bottled-water industry made heroic efforts to provide emergency water supplies in the wake of Katrina and other hurricanes, much of it for free. That made it easy for the powerful beverage lobby to talk lawmakers out of the tax. Even in Michigan, the bottled-water industry prevailed: the Great Lakes' new water compact banned any and all outside diversions. But it gave an exception to the bottlers.

You had to hand it to the bottled-water industry. It nailed a problem that economists from the time of Adam Smith had been noodling over for more than two hundred years. The industry convinced consumers to value water. Now, if only they could be convinced to value water outside the bottle—the water in the ground, in the rivers, and in the tap—America's water woes would be solved overnight. The possibility does not seem imminent.

9

Priceless

BELLE GLADE, FLORIDA, the setting for *Harvest of Shame,* Edward R. Murrow's famed 1960 documentary about migrant farmworkers, has changed all too little in the ensuing decades. It has received international media attention two times since: in the 1980s, when it was named the AIDS capital of the world for its high per capita rate of infection, and again in 2003, when a hanging death brought reporters from around the globe to look into a possible lynching. It was a suicide. But those reporters found plenty of stories to tell—of miserable migrant conditions, of broken race relations.

Belle Glade lies on the far west side of Palm Beach County, surrounded by smoky sugarcane fields and flat winter vegetable farms. On the far east side of the same county is a city that might as well be a million miles away. Snuggled between the azure-blue Atlantic Ocean and a lagoon called Lake Worth, the barrier island of Palm Beach has been winter playground to the nation's wealthy since the 1890s, when Henry Morrison Flagler built his Royal Poinciana Hotel and brought his Florida East Coast Railroad there.

Some of the most spectacular mansions in the United States are hidden behind thick stands of palm and lush tropical foliage along AIA in

Palm Beach; Mediterranean compounds where the swimming pool edges are designed to look as if they are vanishing into the Atlantic Ocean. The pools, elaborate fountains, and, most of all, landscaping make the millionaires who live here—or do not live here, as the case may be in summer—some of the heaviest residential water users in the state. Their average daily use is 13,000 gallons.[1] Back in Belle Glade, the residents, a third of whom live below the federal poverty level, use a lot less: an average of 1,000 gallons a day per household.

So who do you think pays more for water? Answer: the people of Belle Glade. Families pay a flat rate of $62.50 a month for water and sewer service, no matter how much they use.[2] The rate is on the high side of Florida water bills statewide. Residents of the city of Palm Beach have bills closer to the state average: they pay a little less than $30 a month for water and sewage.[3] They are also charged based on the amount they use, so people who use less pay less.

When it comes to water, Palm Beach County is like the globe in miniature. Worldwide, the poor generally pay much more for water than the rich. In a different way, Palm Beach County is a microcosm of the United States as a whole. Across the country, water is priced irrationally, and often inequitably. Most of all, it is priced too low. As a result, Americans do not think about turning off the tap the way they edge up the thermostat on summer days to save a few bucks.

Through general tax revenue, we taxpayers buy the infrastructure that treats water to meet federal drinking-water standards and makes it taste good, too. But these costs are not reflected in our water bill, usually the least painful check we write each month. This is how Americans end up using treated drinking water to flush toilets, wash clothes, and water lawns. More than half of all home water use in the United States goes to keeping grass green. About 14 percent of it is never used at all but leaks out of our pipes.[4]

"This is the only country that you can travel on all the compass points to any city you want, turn the spigot on and get a glass of water, drink it and have a very, very high assurance of safe, high-quality drinking water," says Dr. Ron Linsky of the National Water Research Institute. "But it is the cheapest natural resource in America for the highest quality in the world. The under-pricing of this resource has led to the under-valuing of water."[5]

In the third world, people often pay more for water because they do not have the municipal systems to bring it to them. Some must buy it by the liter. The World Commission on Water for the 21st Century found that people in developing countries pay an average twelve times more per liter of water than fellow citizens who are hooked up to municipal systems. In some cities, the poor pay huge premiums to water vendors over the standard price to those on municipal systems: sixty times more in Jakarta, Indonesia; eighty-three times more in Karachi, Pakistan; and one hundred times more in Port-au-Prince, Haiti. In slums around many cities, water accounts for a big slice of household expenses: 20 percent in Port-au-Prince, for example. "It is stunning that the poor pay more than 10 times as much for water as the rich do, and get poor-quality water to boot," says Ismail Serageldin, the commission's chairman. "A direct link exists between this lack of access and a host of diseases that attack the poor in developing countries."[6]

Some 1.1 billion people around the world lack access to safe water—a number feared to grow to between 2.6 billion and 3.1 billion by 2025. The United Nations attributes 2.2 million deaths a year to poor water and sanitation. The specifics range from diarrhea, in developing countries the cause of 15 percent of all deaths of children under five, to heightened incidence of cholera, typhoid, and viral hepatitis.[7]

The point is that clean water is extraordinarily valuable, the single most important necessity to human life. And people are obviously willing to pay for it. Yet while people in some third-world slums pay up to a quarter of their income for water,[8] most Americans spend less than 1 percent of household income for water.[9] North Americans enjoy not only the highest-treated but by far the cheapest water on the planet. The average U.S. consumer pays $2.30 per thousand gallons. That is about $20 a month for an American family. Among all industrialized nations, only Canadians, with their vast surface-water resources and relatively small population, pay less than Americans. Germans pay the highest price for water among industrialized countries, at a little over $8 per thousand gallons.[10]

In the United States, the price of water is based upon politics rather than economics. Local elected officials help keep water cheap in places like Palm Beach—even in arid cities such as Las Vegas. Vegas residents pay just $2 per thousand gallons. That is less than average for the United

States as a whole despite the overwhelming effort it takes to maintain that oasis in the desert.[11]

So *why* is water cheap—not just in the United States but across the industrialized world? Those of us who are not economists remember Adam Smith, father of the discipline, as the invisible hand guy. The eighteenth-century Scottish philosopher created his intellectual framework for the free market with the metaphor of an invisible hand, showing how self-interest can guide the most efficient use of resources in a nation's economy to result in overall public good.

But when it comes to water, Smith's laissez-faire pronouncements don't float. Smith said as much in *The Wealth of Nations* in 1776, when he pondered a problem still argued over by economists. It is called the Water-Diamond Paradox. Smith put it this way: "The things which have the greatest value in use have frequently little or no value in exchange; and, on the contrary, those which have the greatest value in exchange have frequently little or no value in use. Nothing is more useful than water: but it will purchase scarce any thing; scarce any thing can be had in exchange for it. A diamond, on the contrary, has scarce any value in use; but a very great quantity of other goods may frequently be had in exchange for it."[12]

In other words, why do we give our most valuable resource away for practically nothing and pay gobs of money for one that has no practical purpose? The answer, of course, is relative scarcity. It is just like when people grouse that society values professional baseball players over schoolteachers. The fact is, there are about 4.4 million teachers in the nation.[13] But there are only about ten guys in Major League Baseball who can hit forty or more home runs a year.

While water may seem both plentiful and free, pumping it, treating it, and piping it around are all expensive endeavors. At the Yale School of Forestry and Environmental Studies, the environmental economist Sheila Cavanagh Olmstead throws out two more biggies that most of us do not think about when we turn on the tap: long-run marginal costs and opportunity costs. The marginal costs refer to the long-term water supply expenses that will result from today's decisions to consume more water. Every new housing development, every new golf course, every new tree planted will have some impact on future water-supply needs. "As long as those long-run issues are not folded into the current price of water, no one is going to understand the true value of the resource," Olmstead says.[14]

New sources of water required to meet these increasing demands cost more to develop and transport than older sources. Desalinated water, for example, costs about four times more than groundwater to deliver to consumers. Yet most utilities price water by averaging low-cost older sources with high-cost new facilities. This calculation ensures consumers will never pay the true costs of water.

Besides not taking expensive new facilities into full account, local governments often ignore another key cost when pricing water. The bulk water we give away to farmers, water bottlers, or anyone else has alternative uses that ought to be taken into account. But these resource "opportunity costs" usually go unconsidered. Think of these costs as lost opportunities: The idea that every gallon used is one not used to produce electricity through hydropower, for example, or not available for wetlands restoration or environmental preservation or to meet some other value to society.

In the United States, the explosion of federal water projects in the twentieth century created the illusion that individuals and local governments did not have to worry about the effort or costs of providing water. In the West, since the federal government picked up the tab for the Bureau of Reclamation's more than 600 dams and reservoirs and 55,000 miles of irrigation canals and conduits, tens of millions of people considered it perfectly rational to live, work, and farm in what should have been the most inhospitable land in the country. In much the same way, Olmstead says, pouring thousands of gallons of water on your lawn in arid Arizona "is a perfectly rational decision on the part of homeowners because water prices are so low."

The political pork barrel rolls on. Members of Congress from water-stressed states are forging coalitions to make sure the federal government subsidizes desalination plants all over the nation, the subject of more attention in chapter 11. This time around, the subsidies will make it seem perfectly rational to build more golf course communities in California, Texas, and Florida. But what is good for retiring baby boomers who like a round of eighteen holes will not be so for their children and grandchildren. If the current generation of Americans refuses to pay the true cost of water, future generations will face not only substantially higher prices but significant interest on the growing national debt.

Following the great dam-building era of the early twentieth century, the next wave of water-infrastructure subsidies came in the 1970s after passage of the Clean Water Act and the Safe Drinking Water Act. The federal government doled out billions of dollars to local communities to help upgrade infrastructure to comply with the tougher standards. That trend slowed during the Reagan Revolution when the feds tried to shift costs back to local utilities. Instead, the utilities simply did not do as many upgrades. Now, the pipes that distribute clean water and collect wastewater in most cities and towns have passed their life expectancy. The result is a nearly $1 trillion bill coming due for critical drinking-water and wastewater investment in the United States over the next twenty years.[15] Construction and repair of water and wastewater systems is now the number one infrastructure need in the United States.

"We can turn on faucets at any time of the day or night and expect clean water," says G. Tracy Mehan III, former assistant administrator of the EPA's Office of Water. "Monthly water bills, for most of us, hardly approach the cost of cable TV. But underneath this rosy picture lies a monster."[16]

Australians call the monster the Nessie Curve. They named it after the Loch Ness Monster because so much of it, just like water pipes, lies beneath the surface. The same demographics that created the looming future shortfalls for Social Security, Mehan says, are creating a similar liability for the nation's water systems. Thousands of miles of pipe laid more than one hundred years ago must be replaced in the next couple decades. Treatment plants have a much shorter life of between twenty-five and forty years. They also will have to be replaced or overhauled in the coming years to meet EPA standards.[17]

Yet few politicians talk about increasing the price of water to pay for these investments; they do not win votes by signaling citizens to turn off the spigots through higher prices or user fees. Instead, reelection requires that they bring home the bacon to local voters, landing water projects that make constituents think they are getting something for nothing. The costs, of course, will be passed on to future taxpayers.

Americans will plunk down a thousand times more for a pint bottle of water at the corner 7-Eleven than they pay for their tap water, even though there is scant difference between the two. Yet we fend off local governments' attempts to raise rates for water and sewage. Somehow, we value the water in the plastic but not from the tap. Why? Americans have been brought up to believe that their water is a right, a very low-cost one, rather than a commodity. And they do not take kindly to public officials who would try to charge them what it truly costs. Public outcry often convinces politicians to defer infrastructure investment, maintaining shabbier and shabbier systems on fewer dollars.

Lots of local elected officials can recount the story of the hapless Tucson, Arizona, City Council: In the late 1970s, following a severe drought, the city of Tucson became the first in the nation to adopt marginal cost rates for water; that is, the city wove the costs of additional water into its price. One year after adopting those rates, the entire city council was voted out of office.[18] When stewardship is rewarded this way, it is easy to see how it has become the exception rather than the rule.

A more visible monster than the Nessie Curve is the threat of terrorism to the nation's water supply. It is also a costly one. In his State of the Union Address in 2002, President Bush warned Americans that U.S. soldiers fighting in Afghanistan had found diagrams of U.S. public water facilities.[19] That same year, Congress passed the Bioterrorism Act that required drinking-water utilities to conduct "vulnerability assessments" and take a harder look at emergency response plans. The resulting reports exposed significant security needs, from small things like fixing fences to biggies like relocating pipelines and distribution mains. The most critical, early fixes are estimated at $1.6 billion nationwide, according to the American Water Works Association.[20]

But again, no one dares argue that local consumers should pay this bill through the issuance of bonds; that would require us to face costs directly. Elected officials instead will hide costs in national legislation and reallocate them, predicts Sanford V. Berg, director of Water Studies at the University of Florida's Public Utility Research Center. This distortion, he says, will result in higher water costs overall. "When we subsidize capital investments and not operating costs, utilities will tend to over-invest in facilities," says Berg, "because someone else is footing the bill."[21]

It has a familiar ring, doesn't it? Discussed in chapter 3, this was the rationale used to drum up local support for the Animas–La Plata dam now underway in Colorado: "WHY WE SHOULD SUPPORT THE ANIMAS-LA PLATA PROJECT: BECAUSE SOMEONE ELSE IS PAYING MOST OF THE TAB! We get the water. We get the reservoir. They pay the bill."

During the worst of Florida's last drought, when the Council of 100 released its controversial report on water supply, other organizations such as the Florida Chamber of Commerce quickly followed with recommendations of their own. Among the ideas, from desalination to water transfers to a statewide water board, not one word was uttered about the cheap price of water. "It's an absolute deal killer," says Berg. "The most obvious, powerful and direct way to manage the water problem would be to pay for new investments through higher prices.

"Politically, we seem to believe there is a free lunch, or at least a lunch that someone else will pay for," he says. "Such 'mathemagic' is a recipe for a water crisis in the not-too-distant future."[22]

Used to be, advocating true-cost pricing for water was the job of conservatives over at the Cato Institute and other right-leaning think tanks. Traditionally, people looked at pricing in one of two ways: Economists generally argued that water should be treated as an economic good; users should pay the costs of the service, an incentive to conserve. The human-rights and environmental communities, on the other hand, argued that water is a basic necessity and a human right, and that it should not be viewed as an economic commodity. But arguing one way or the other does not make sense because both sides are right. If you do not price water, people are going to waste it. Yet citizens also have a right to clean water.

As it becomes increasingly obvious that users will waste free water until it is gone, environmentalists increasingly view water through an economic lens. Users will waste free water even against their own interest, as in the case of the Florida developers who in some areas drained water until there was none left for new growth. Previous chapters detail how Florida and other eastern states hand over millions of gallons of water a day to most anyone who wants it: farmers who then may waste it on inefficient flood irrigation, bottled-water companies that turn it into a commodity and sell it for a dollar a liter. So while a watermelon in the supermarket bin has value, and a bottle of Evian on the restaurant menu has value, the water in our aquifers, rivers, and streams has virtually none. Groundwater and surface water, when no one has to pay for them, will be abused until no one can use them.

This inevitability is known as the Tragedy of the Commons. In his essay by that name in 1968, biology professor Garrett Hardin wove the story of a village with a commons where herdsmen graze their cattle. There are no rules, but each herdsman decides how many cattle to graze.

It is to the advantage of each herdsman to keep increasing the size of his herd, even when it becomes clear that the pasture is being overgrazed. "Therein is the tragedy," Hardin wrote. "Each man is locked into a system that compels him to increase his herd without limit—in a world that is limited. Ruin is the destination toward which all men rush, each pursuing his own best interest in a society that believes in the freedom of the commons. Freedom in a commons brings ruin to all."[23]

The Ipswich River watershed in northeastern Massachusetts could be Hardin's commons. Covering a 155-square-mile area with 22 towns, the watershed powered the region's economic development, giving water, sustenance, power, and transportation to early settlers and their descendants. Its abundant fisheries and famous clam beds were the shimmering pearls of the economy, even as, in the early twentieth century, the communities of the Massachusetts North Shore increasingly merged into urban Boston. Fueled by commuter rail lines and highways, the North Shore began to sprawl with big, thirsty housing developments that used, because of their lawns and swimming pools, 50 percent more water than traditional developments in the watershed.[24]

Water supply should not be a problem in this region, blessed with an abundant average forty inches of rainfall each year. Yet communities along the Ipswich River have overpumped groundwater to the extent that, over the past decade, parts of the river have gone completely dry each summer. The river dries up when the lawn sprinklers whir and the pools fill, even though average withdrawals, some 30 million gallons a day, are well within the amount permitted by the Massachusetts Department of Environmental Protection. In other words, state environmental regulators have handed local utilities more water than exists in the river to supply 330,000 people and thousands of businesses. This "almost exclusive focus on meeting the growing consumptive demand for water, while ignoring the other functions and values of the river," has led to a crisis with dire consequences for the ecosystem, the shellfish industry, and recreation.[25] About a third of the clam beds for which the Ipswich is so well known are closed at any given time. Fish kills are becoming more common, causing a clear environmental loss as well as a stink. Kayaking, canoeing, fishing, and other ecotourism pursuits dry up, obviously, when the river does.[26]

Kerry Mackin, executive director of the Ipswich River Watershed

Association (IRWA), argues an underlying cause of the Ipswich crisis is that permit holders pay nothing to withdraw water from the basin. In 2003, the IRWA and others filed suit against the Massachusetts Department of Environmental Protection. They charged the department's overpermitting violated its own regulations, causing "drastically reduced stream flows in the Ipswich River resulting in fish kills, habitat destruction and severe ecosystem impairment." The IRWA wants the state to charge utilities for withdrawals to better reflect the true value of water, encourage more efficient use of it, and raise money for conservation and other programs. Toward those ends, the organization conducted a resource economics study of the river that asked residents about their willingness to pay more for water. Respondents were willing to pay an average $31 more per household a year to "restore flows needed to sustain healthy fish populations."

The Tampa Bay Water Wars were another classic example of the Tragedy of the Commons. Local officials sucked up as much water as they could for their own constituents and turned a blind eye to the larger consequences. Not only did they contaminate their own groundwater but they dried up wetlands, triggered sinkholes, and drained lakes. Their actions devastated local ecosystems as well as property values.[27] During the Tampa Bay crisis, the Florida legislature cracked down on the state's water-management districts and made them figure out the "minimum flows and levels" that must stay in-stream to sustain water supply. The minimum-flows requirement had been on the books since 1972, but the water districts had never tackled the massive calculation job. When they finally did, some districts saw that they had overpermitted groundwater use by millions of gallons a day.

In Tampa Bay, residents whose wells went dry and whose lake beds turned to weeds have come to value water. So have out-of-work Massachusetts clammers and fly fishermen who can no longer catch native trout in the Ipswich River. It sometimes takes a crisis for Americans to appreciate their readily available water.

In 2005, heavy rains raised the turbidity of drinking water in Phoenix, and the city issued a boil-water notice as a precaution. For the first time—and for just one day—1.4 million people were without a service they took for granted. Jokes went around about the crisis of a day without a latte. "Welcome to Phoenix," deadpanned one editorial writer, "where our library computers are filtered better than our drinking water."[28] Guffaws aside, the one-day alert was a real hardship for some. Restaurants shut down. Hospitals had to postpone surgeries.[29]

In the fall of 2003, during the huge power blackout that stunned the northeastern and midwestern United States, Cleveland and Detroit were among the major cities whose pumping stations lost power, cutting off drinking-water service as well as water pressure for fighting fires. The stress of bringing the aging water systems back on-line doubled the number of water-main breaks in the cities. The blackout also caused major sewage spills into waterways in Cleveland and New York.[30]

Residents of Milwaukee, Wisconsin, know the value of clean water better than most Americans. In 1993, the community had an outbreak of cryptosporidiosis, a parasite that causes gastrointestinal illness that can be transmitted through water. More than 100 people died; some 400,000 became sick. Says Kathryn "Katie" McCain, a past president of the American Water Works Association who put in three decades with Dallas Water Utilities: "It was a tragic reminder that the work water suppliers do every day has life and death consequences."[31]

It does not make sense to wait for a terrorist attack before spending money on emergency generators to pump water, or to wait for bacterial outbreak before upgrading filtration. The price elasticity of demand measures how consumers respond to changes in price. Opponents of water-rate increases often argue that water demand is price inelastic—in other words, that consumers do not respond when rates go up because water is a basic necessity. But several recent studies show this is not true. In 2005, Florida's water-management districts funded a sixteen-community study on how rates affect single-family residential water use. It made a clear statistical case that price increases lead to lower use. The study found that in homes with access to alternative sources, such as reclaimed water, increasing the price from $1.20 to $2 for every thousand gallons reduced per capita use from 140 gallons a day to 116 gallons a day, or 17 percent. For families that did not have access to alternative water supplies, the same price increase saw per capita use drop from 161 gallons a day to 140 gallons a day, or 13 percent.[32]

Advocates for the poor worry that higher water rates place an undue burden on low-income residents. Berg argues that since we are already subsidizing well-off Americans with cheap water, it would make more sense to subsidize water for the poor and make everyone else pay the true cost of the resource. "Targeted subsidies make much more sense than sending uneconomically low price signals to everyone including the wealthy," he says.

In the United States and even in some parts of the world with intractable water-supply problems, such as India, water-pricing reform has begun to spur conservation and set aside more water for natural systems. Price reform does not always mean higher rates; it can mean lower rates for smaller blocks of water use and higher rates for bigger blocks. It also should not be aimed solely at utility customers. Pricing water properly may be most important in agriculture, where water subsidies are larger and more pervasive than in any other realm of water use.[33] In the West, Imperial Valley farmers pay $15 an acre-foot for the same Colorado River water that costs urban users in Los Angeles $400 an acre-foot. And if farmers have to pay more, the newer, plastic commodities should have to pitch in, too: states hand groundwater to bottled-water companies for free, and the companies then charge consumers a thousand times more than the cost of virtually identical tap water.

Will water become the oil of the twenty-first century? This was the question, in recent years, on the minds of everyone from market speculators to U.S. intelligence agents. Peter Gleick of the Pacific Institute gave the old question a new spin: "Are we going to permit water to become a commodity like oil, to be over-pumped and under-priced, and used wastefully, leading to water wars, international conflict and competition and environmental destruction?" Or could we learn from the lessons of oil dependence and make sure water "is used efficiently and allocated properly, through national policies and international cooperation, and that the environment is protected from damage caused by its extraction?"[34]

What better place to hunt for the answers than the oil-rich, water-poor state of Texas? There, the nation's most infamous oil speculator had turned his attention to a brand-new resource. No one could accuse Boone Pickens of undervaluing water. He wanted to set its price tag nice and high.

10

Water Wildcatters

IN THE GRASSY HILLS of northwest Texas, a sprawling ranch called Mesa Vista sits atop one of the last untapped parts of the Ogallala Aquifer. The largest aquifer in North America, the Ogallala flows for 174,000 square miles under the Great Plains from South Dakota to the Texas panhandle. It is the source of water for the nation's breadbasket, supplying a third of all the groundwater used for irrigation in the United States. But the Ogallala does not recharge like most aquifers. Once the water is gone, it's gone. The aquifer holds what's known as "fossil water," left from an ice age 10,000 years ago. Its depths range widely from just a foot to 1,300, but the average is 200 feet. When farmers got hold of cheap, deep electric pumps in the 1950s, they began to pull up the ancient water at unsustainable rates. As a result, the Ogallala has dropped more than 100 feet in parts of Kansas, New Mexico, Oklahoma, and Texas.[1]

When you suck up water from one part of the Ogallala, you can dry up other parts, too. And so it is only a matter of time before all the wells sunk in this huge aquifer go dry—the Tragedy of the Commons. This fact is well-known to Boone Pickens, who owns the 27,000-acre Mesa Vista Ranch. A geologist by training who began his career as an oil roughneck in Oklahoma and Texas, Pickens earned fame and fortune with his uncanny understanding of the oil and gas reserves that flow underground. In 1997, when the Canadian River Municipal Water Authority, serving nearby Amarillo and Lubbock, bought water rights to 43,000 acres of land just south of the Mesa Vista, Pickens saw that it would be able to suck up the water under his land, too. With the corporate-raider spirit that gained him notoriety in the 1980s for attempting hostile takeovers of Gulf Oil, Unocal, and Phillips Petroleum, Pickens decided to sell his own water before the authority could drain it. "To just sit there and not do anything would probably border on being stupid," he says.[2]

Pickens formed a company called Mesa Water, and began buying up water rights from his rancher neighbors for $350 to $500 an acre, substantially more than going prices for the land itself. One rancher said he could earn more money on water than Herefords.

"We could all haul hay in our Mercedes," said another.[3]

By 2005, Pickens had cinched water contracts with 75 panhandle landowners, securing 100,000 acres of land at a total cost of $25 million.[4] Now, he has to sit tight until one of the cities downstate becomes desperate enough to buy. The population of San Antonio, for example, is expected to double by 2050, yet the Edwards Aquifer the city relies on for drinking water is tapped out. Mesa's engineers have figured out how they could build a 171-mile pipeline large enough to drive a car through to deliver water from the panhandle. Pickens has secured financing for the $1.5 billion pipeline through tax-exempt bonds sold by J. P. Morgan. He estimates he will someday sell his liquid gold for up to $1,200 an acre-foot.[5]

For now, officials in San Antonio and elsewhere are emphatic that they are not buying. They say they do not want private speculators to wrest control of Texas's water resources. But Pickens can afford to wait. At 77, he has an estimated net worth of $1.5 billion that is soaring along with his energy hedge funds, BP Capital. (Not to mention a wife twenty years his junior.) "I will be here to see water sold and the infrastructure built and the water delivered," he predicts. "I think that will happen in less than ten years."[6]

Like a bowlegged cowboy, the Lone Star State straddles the 100th meridian that divides the moist East from the arid West. Texas also sits smack in the middle of the trend toward private water markets in the

United States. In the West, water rights have been bought and sold like pork bellies for decades, as fast-growing cities pay farmers who hold senior water rights not to farm in dry times. In Texas, if Pickens gets his way, he will be the first private water broker in the state, and surely not the last. Other speculators were lining up to buy up rights too, in hopes they would become rich like the investors who followed Pickens in his corporate raider days.

In the East, water markets were unheard of until fairly recently because no one needed water badly enough to buy it. But markets emerge where scarcity exists. And emerge they have in Florida—even though they are not exactly legal.

For thirty years, John "Woody" Wodraska was a public steward of water, in charge of the South Florida Water Management District before he moved west to take the helm of the Metropolitan Water District of Southern California in 1993. As general manager of the California district, one of the largest wholesalers of drinking water in the nation with 16 million customers, Wodraska had a mantra. The district should run "like a Fortune 500 company," he used to say—innovative, efficient and inexpensive.[7]

In 2000, however, Wodraska came back to Florida to flog just the opposite idea. Now North American director of a publicly traded water company called Azurix, Wodraska's new mission was to try to convince Florida lawmakers that private companies should be given a much larger role in managing and delivering water. Government, he argued, simply could not do it as well. "There's an ethic in managing costs in the private sector that simply does not exist in the public sector," he insisted.

Plenty of private companies run water utilities in Florida. Azurix wanted to do much more. While Florida's water-management districts were historically too generous in doling out water permits, they hold firm control over those permits compared to western states. For one, permits have a time limit, from a few years up to twenty. For another, water permits in Florida are not supposed to be transferred for a different use, much less sold. Unlike places with water markets, such as California, if a Florida farmer has a permit to pump 4 million gallons of water a day and he uses only 2 million, he cannot sell the other 2 million to a golf course.

Azurix lobbied for changes in state law that would create markets for permits to be bought and sold. Wodraska, along with Governor Jeb

Bush's environmental chief David Struhs, pushed the free-market idea with lawmakers and water managers by arguing it would result in cheaper water doled out more efficiently. Letting permits find their value in the market, Wodraska said, also would create incentives for conservation.

Azurix officials knew those arguments would not be enough to convince the Florida legislature to hand a private company control of the state's most important resource. So Azurix came up with a deal as sweet as the sugarcane growing south of Lake Okeechobee. It offered to finance and manage a chunk of the massive Everglades restoration effort in exchange for rights to sell the water it would store as part of the project. Azurix officials were enamored with a technology known as Aquifer Storage and Recovery, which skims excess surface water in the soggiest periods and stores it hundreds of feet underground to pull up later during dry times. Azurix offered to spend billions of dollars on water plants that would feature underground injection wells throughout southeast Florida.

Azurix's pitch looked like this: Take over a daunting organizational challenge requiring big up-front capital. In return, get a long-term license to sell a commodity—in this case a thirty-year monopoly license to sell water to the 7 million residents of South Florida.[8]

It sounded remarkably similar to the one that a fast-growing Houston energy broker called Enron was making in the fields of gas and electricity. That's because Azurix and Enron were one and the same.

Enron created its water spin-off to try to make a splash in what it estimated is a $400 billion global market for water. Instead, Enron's attempts to take over water utilities from Argentina to the United Kingdom and to create water markets in the U.S. Sunbelt resulted in $900 million in debt for Azurix—a factor in Enron's devastating collapse of 2001. The *Sarasota Herald-Tribune* newspaper said it best: "Jeb Bush and the people of Florida might want to thank their lucky stars" that Enron imploded before it managed to snag major public water contracts in Florida.[9]

Regardless of their misdeeds, Enron officials were prescient in their predictions that Texas and Florida would soon open to water markets. Many recoil at the idea of markets for a resource they hold sacred. Maude Barlow, the eloquent chairperson of the Council of Canadians who is fighting the "commodification" of water all over the globe, puts it this way: "there are some areas of life that should be marked a part of 'the commons' and set aside from the rules of the marketplace. Water is one of them."[10]

But making water a part of the commons has led to countless Tragedies of the Commons, even in the most water-rich parts of the country, from the Everglades to the Chesapeake Bay. If Florida's environmental history teaches anything, it is that handing over the state's most precious resources, for free, to anyone with a business plan, does not work. Appropriate pricing, markets for buying and selling water, and other economic incentives for its wise use all "have a central role to play in the transition to an era of scarcity," says Sandra Postel of the Worldwatch Institute, the global sustainability think tank.[11]

In Florida, markets could help move water from low-value uses to higher ones while promoting conservation.[12] The lesson of Azurix's short-lived courtship was that the state, not the private sector, should design such markets and set up a regulatory system to prevent supply and price manipulation. The state would insist, just as it did during negotiations over the Apalachicola-Chattahoochee-Flint Compact, on strict conservation for users and minimum stream flows for fish and wildlife. If they allowed such trading at all, state lawmakers would have to come up with a way to make sure it met the three-part test at the heart of Florida water law: that the proposed use (1) is reasonable-beneficial, (2) does not interfere with existing legal users, and (3) is in the public interest.

But Azurix's splash into Florida so turned off lawmakers and the public—part of the problem was the company's secrecy—that it killed any chance of a state water market for the following decade. So instead of getting ahead of the trend, the Florida legislature got left behind. Today, Florida has a water market. It is a gray market over which lawmakers have no control.

Along a 700-square-mile strip of Sarasota, Manatee, and Hillsborough counties, groundwater pumping caused such severe saltwater intrusion and precipitous dips in the aquifer that in the early 1990s, water managers said they would grant no new groundwater permits there. But the force of growth along this stretch of Florida is as powerful as a hurricane, and about as easy to stop. Over time, the water managers and the water users evolved a permit-trading system that reduces groundwater pumping and gets water to those who need it. The users are actually buying and selling the permits—a fact the water managers know about but ignore.

The deals look like this: Existing permit holders give up part of their groundwater allocation in exchange for approval to transfer some of their gallons to a new user. Say a farmer withdraws 2 million gallons a day. Water brokers may work out a deal where the farmer cuts his use to 500,000 gallons a day, 500,000 more go back to the aquifer, and the farmer sells his rights to the other million to a golf course community that needs water. Regulators go for the deals because they reduce overall groundwater pumping. "Whether or not there's a financial arrangement between the two parties is not relevant to us," says Michael Molligan, spokesman for the Southwest Florida Water Management District. "What we're trying to do is ensure a net benefit for the resource."[13]

But the financial side is significant, according to brokers, ranging from a onetime $100,000 payment to multiyear contracts worth millions. From a conservation standpoint, the deals may be in the public interest. But an unregulated market in which water permits are bought and sold without oversight is not. Florida has some of the strongest open-government laws in the nation: the so-called Sunshine Laws that say all government business must be conducted in the open. If Florida lawmakers were going to set up water markets, presumably they would provide for public disclosure and establish priorities above and beyond who has the biggest wad of cash. Southwest Florida's market is open only to those who hire the lawyers and other consultants who know about it. No one is privy to price. And there seems to be no criteria but price. What if two new users, say, a tomato farmer and a home builder, wanted to bid for the same gallons? An open process might reward the one who promises the biggest conservation gains or sets aside the biggest chunk of wetlands, rather than the one who writes the biggest check.

"Right now people may be ignoring it, but markets always find a way," says Mark D. Farrell, an engineer and principal at Water Resource Associates of Tampa, one of the companies brokering the transfers. "I think you can have the best of both worlds. As long as you start with the premise that the resource is protected first, then you can let economics do its thing."[14]

The Swedish privatization advocate Fredrik Segerfeldt found that similar informal trading underway in India led to tensions and efficiency losses. "The government, perceiving its advantages, has opted for nonintervention and has turned a blind eye," Segerfeldt says. "But the informality of the trade means that there is no one to ensure that agreements are adhered to." In Pakistan, the government legalized spontaneous trading that was going on in 70 percent of water courses in one study, which led to a 40 percent increase in the price farmers could demand for water rights.[15]

Water markets are just one-half of the water-privatization trend that detractors fear is sweeping the globe. The other half is privatization of water utilities. Here is the problem with opposing either one of them: both are well underway whether we like it or not. Where scarcity exists, markets emerge. This has proven true even in Florida, which has never sanctioned water markets. Meanwhile private companies run about a fifth of water systems in the United States. So the choice is not to be for privatization or against it. Rather, the choice is to ignore it or to regulate it smartly to ensure water for both people and the planet. Local elected officials across the nation may contract with private utilities to provide water if that is in the best interest of their citizens and resources. Around the world, water privatization has its horror stories and its successes. There are lessons in all of them.

The death of a seventeen-year-old boy named Victor Hugo Daza in Cochabamba, Bolivia, in 2000 underscores how volatile the water-privatization issue has become worldwide. Daza was shot by police in a citizen uprising after a Bechtel subsidiary called Aguas del Tunari landed a contract so unfair to the local people that the company could charge them for water they collected from their own rain barrels. The World Bank had insisted Bolivia turn over its public water utility to the private sector, or the World Bank would refuse to guarantee a $25 million loan for infrastructure improvements. Bechtel and a British consortium of investors put up less than $20,000 up-front for the multimillion-dollar system. The company immediately raised the price of water beyond the reach of the majority of the population, and expected to earn an annual income of $58 million. A five-foot-tall mechanic named Oscar Olivera led citizens to strike, and came under death threats from the military. Thousands took to the streets in a confrontation with the army that left many injured and six, including Daza, dead. Bechtel was forced out of town. Cochabamba's water services now are run by a citizen-controlled, nonprofit company.[16]

Over the past twenty years, water privatization has spread rapidly in poorer countries such as Bolivia as the World Bank made governments privatize utilities in exchange for loans. While private companies still run only 5 percent of the world's waterworks, their recent growth has been significant. In 1990, about 51 million people got their water from private companies. By 2002, that number had increased sixfold, to 300 million.[17]

Two of the three biggest water firms in the world are French: Suez and Veolia Environment, spun out of Vivendi in 2002. They have dominated water supply in France since the nineteenth century and seemed poised, at the turn of the twenty-first, to take over waterworks around the world. In the late 1990s and early 2000s, the French firms, along with German energy giant RWE, went on a buying spree of the largest U.S. private water utilities. Vivendi paid more than $6 billion for US Filter in 1999. Suez paid more than $1 billion for United Water Resources in 2000. RWE, in 2003, paid $4.6 billion for American Water, the largest private provider in the United States.

An investigation by the International Consortium of Investigative Journalists warned, "Having firmly established themselves in Europe, Africa, Latin America and Asia, the water companies are expanding into the far more lucrative market of the United States." An RWE executive told the journalists that within the decade the company expected to double its market to 150 million customers, primarily because of U.S. expansion.[18]

But don't rush to call your stockbroker just yet. The U.S. water market may not be nearly as lucrative as some thought, at least for the multinational firms. In 2004, Veolia shed most of US Filter to Siemens at an overall loss of $4 billion.[19] In late 2005, RWE officials announced they were putting all North American and UK water interests up for sale, including American Water and Thames Water, famously privatized by former British prime minister Margaret Thatcher. The concept of a global water business had "not really worked," RWE chief executive Harry Roels admitted to the *Financial Times.* "Scale and synergy effects in the water business are regional," he said, "not global."[20]

Suez seems to have stumbled the hardest in the U.S. market, or at least the most visibly. In 2003, the largest water-privatization deal in U.S. history, between Atlanta and United Water, fell apart amid citizen outrage over dry fire hydrants, slow service, and brown water. "The city had a motto for years, and it went something like 'Atlanta grows where water goes,'" Jack Ravan, Atlanta commissioner of watershed management, told the *New York Times.* "I think we've learned enough to know that we'd prefer to see the city in charge of that destiny."[21]

Atlanta had good reasons for entering a twenty-year contract with United Water to privatize the city's water utility in 1999. Like so many cities across the nation, Atlanta faced steep costs to repair aging sewer pipes, treatment plants, and other infrastructure. United Water's $22 million a year contract was $20 million less than Atlanta spent for its public utility. The savings could buy capital improvements.[22]

In hindsight, the deal sounded too good to be true. And it was. United Water officials said the city's water-infrastructure problems were much greater than they had been led to believe. The company said it was losing $10 million a year in a bad contract that the city refused to renegotiate. Citizens, meanwhile, endured numerous violations of drinking-water standards and a string of maintenance problems—from open manholes to water-main leaks that went unrepaired for weeks. In fall 2002, Mayor Shirley Franklin announced, "United Water has not lived up to its responsibility." She gave the company ninety days to fix the problems or lose the contract.[23] The following January, both sides agreed to nix it.

The Atlanta case provides good ammo for privatization critics who say private water companies often promise the moon to land contracts and then try to renegotiate them later. Critics also charge that private utilities are less accountable, less vigilant about water quality, and more expensive than public utilities. However, a far-reaching investigation by the AEI-Brookings Joint Center for Regulatory Studies found that in the United States private customers do not appear to pay more for water. The study, which included every community water system in the nation and covered the years 1997 to 2003, found, in fact, household expenditures on water decreased slightly as the local share of private ownership increased. The study found that privately owned systems, on average, complied with drinking-water regulations just as well, in some cases better, than publicly owned ones. The privates had slightly higher monitoring and reporting violations.[24]

Overall, the results said more about the potential for benchmarking and competition than whether public or private utilities are superior. The study found that the higher the share of a local population served by a single system, the less likely that system is to be in compliance with the Safe Drinking Water Act. "Overall, the results suggest that absent competition, whether water systems are owned by private firms or governments may, on average, simply not matter much," the authors concluded.[25]

It is unusual to find such middle ground in the debate over privatization. Advocates say governments, especially in the third world, have proven their incompetence to provide clean water, evidenced by the fact that a sixth of the people on the planet do not have access to it. In the United States, they say, profit motives give corporations strong incentives to conserve, and to see to it that their customers are served rather than water being spilled.[26] Groups such as Public Citizen warn, on the other hand, "there are more than enough cases that expose the opposite side," and that privatization efforts could backfire on taxpayers.[27]

The fact is private companies have both improved and botched water services in the United States and beyond. And so have governments. Private companies have a role in helping solve the global water crises. They also have a role to provide water supply to U.S. cities that are strapped for cash and need new investments to upgrade aging systems.

Aqua America, the largest publicly traded water utility in the United States, has a growth strategy of buying up smaller systems. The company bought more than sixty waterworks in Florida in 2004, a transaction that so far has proven smooth for consumers. Economies of scale mean the company can cover upgrades without sharply raising rates, says Chairman Nick DeBenedictis. The Pennsylvania-based company, which serves 2.5 million people nationwide, sees enormous growth potential across the Sunbelt in serving new housing developments that existing utilities cannot accommodate.

Analysts predict more and more water managers will look to private industry for access to capital for the $1 trillion bill coming due for those leaky, decades-old pipes and pumps. (DeBenedictis says his company, for one, is steering clear of older cities and decrepit pipes.) Meanwhile local politicians are often happy to contract with private companies because it lets them off the hook for difficult decisions like raising rates. In 1997, about 400 governments in the United States had entered into long-term contracts with private companies to run their waterworks. Five years later, that number was 1,100.[28]

The trick for local governments is to get past the two extreme views of privatization to figure out potential providers' financial health, experience with similar-sized systems, customer-service record, and commitment to conservation. "The rhetoric will inevitably lead to bad public policy," says Tony Arnold, a property and land-use law professor at the University of Louisville's Brandeis School of Law. "The more important issues involve identifying under what conditions water privatization should occur and what safeguards and accountability mechanisms should be provided to protect the public."[29]

Utility experts suggest that benchmarking offers the best tool for governments, investors, and consumers to judge a water utility, whether public or private. Financial performance, water quality, customer service, and other measures pulled together in a ranking prove invaluable whether decision makers are in the third world or the United States.[30] The California-based Pacific Institute has drawn praise for its thirteen principles for privatization that combine social and economic objectives. They include meeting basic human needs, meeting ecosystem needs, providing subsidies when necessary to overcome poverty, fair and reasonable water rates, linking rate increases to agreed-upon improvements, using subsidies only when economically sound, and making water companies show that new supply projects are less expensive than conservation projects. The objectives also address the responsibilities on the government's side: public water-quality monitoring, high-quality contracts, clear dispute-resolution procedures, independent technical review, transparency during contract negotiations, and, perhaps most importantly: retaining public ownership and control over the actual water.[31]

This last point has not been lost on Woody Wodraska, the former North American director of Enron's Azurix. Still smarting from what he calls "the black part of my résumé," Wodraska has bounced back and is today national director of water resources for engineering giant PBS&J. Instead of selling water, PBS&J sells consulting hours, in high demand from both public and private water utilities.

If he has come to believe one thing, Wodraska says, it is that "you have to start by protecting society's interests first." The public should always own the water, whether private companies pipe it around, whether its permits are bought and sold. "I come down on the side that water should be a public resource and protected," Wodraska says. "But that doesn't mean you can't manage it more efficiently."

After working on both sides, Wodraska came to the conclusion that the world's water woes will not be solved without significant help from both governments and private companies.

They will need plenty of help from consultants, of course. That could get pretty expensive. Especially when it comes to costly, yet unproven, new technologies.

11

Technology's Promise

KINGS HIGHWAY in southwest Florida stretches from Charlotte Harbor northeast into palmetto scrub and cattle ranches. Until recently, it was a lonely two-lane road with little moving but the solitary red-tailed hawk overhead, the dim-witted armadillo just missing your wheel. Pale blue sky and native grasses made for a wide view, broken only by barbed wire and the occasional makeshift roadside memorial.

The serpentine highway was named, appropriately, for the man considered the "Cattle King" of South Florida, a six-foot-six-tall rancher named Ziba King. But in recent years, the cattle ranches along Kings Highway have fallen one by one. The widened highway now snakes through an entirely different sort of crop: rooftops. Strip shopping centers, banks, and, most of all, middle-class subdivisions fill the horizon. Five thousand new homes are in various phases of construction: concrete slabs here, frames there, real estate signs everywhere. A Super Wal-Mart with gas station and garden center is underway on a 50-acre tract of grassland. Developers are making room, too, for more of the travelers who stream by on Interstate 75—the highway that runs north and south from Detroit, Michigan, to Naples before jogging east across the Everglades. A

La Quinta Inn is going up on one side of a Cracker Barrel, a Sleep Inn on the other.

The development pattern is both familiar and relentless. As Florida's coastline fills up, urbanization marches inland. The farther from the beach, the more affordable the single-family home, the cheaper the hotel chain. Hurricane Charley devastated this corridor in 2004, leaving 10,000 people homeless. Two years later, more than 700 families still lived in temporary housing that residents call "FEMA City," a stark, treeless trailer park adjacent to the interstate. But neither Charley nor the three other hurricanes that walloped the area the same year could slow the moving vans. In fact, the devastating 2004 hurricane season had almost no effect on the following year's population growth. "Florida's population grew by more than 400,000" after the quadruple hurricanes, says Stanley K. Smith at the University of Florida's Bureau of Economic and Business Research. "This is one of the largest increases in Florida's history."[1]

Over the coming decade, Smith's demographers predict, Florida will grow another 21 percent, exceeding 21 million people to become the third-largest state behind California and Texas. Between those residents and upward of 70 million tourists a year, regulators predict total demand for water in the state will climb to 9.3 billion gallons a day.[2] That is a billion gallons more than Floridians use now. With groundwater reserves tapped out in so much of the state, politicians and water managers are desperate for new ways to get water to the sprouting subdivisions. They are turning, increasingly, to high-tech alternatives, such as underground storage wells or saltwater desalination.

This drive to find "new" sources of water is underway across the United States, even in slower-growing states such as Wisconsin where aquifers are overdrawn. The search for new water is no longer solely an urgency of the American West. U.S. senators from both sides of the 100th meridian have forged a coalition to make sure more federal funds are spent on water-supply research and technology.[3] "Despite receiving substantially more rainfall than the western United States, much of the east coast is facing water shortages," U.S. Senator Pete V. Domenici, New Mexico Republican and then-chairman of the Senate Energy and Natural Resources Committee, said as he opened a hearing on new water-desalination facilities in 2005. "Boston, Atlanta and much of

Florida are nearing the end of readily available water. Without significant technological advancements that allow us to better utilize, conserve and produce additional water in a cost-effective manner, it is unclear how we will meet this need."[4]

The emphasis is clearly on producing additional water. Much like they championed dam building in the twentieth century, the Bureau of Reclamation and other federal agencies now push expensive supply-side solutions to solve the nation's water woes. The bureau's "Desalination and Water Purification Technology Roadmap," a joint effort with Sandia National Laboratories, assumes that the United States will require 16 trillion additional gallons of water a year by 2020 for municipal and light industrial use—the equivalent of a quarter of the total outflow of all the Great Lakes.[5] This ignores the nation's decreasing per capita use of water, and the fact that industries, from steel factories to farming, use less and less water as efficiencies improve. In the 1930s and 1940s, for example, it took as much as 300 tons of water to produce one ton of steel. By the 1980s, overall water consumption to produce steel had dropped to between 20 and 30 tons.[6]

The government's roadmap notes that "water saved by conservation programs will be insufficient to slake the nation's ever-increasing thirst." It goes on to say that "conservation activities can also reduce the volume of in-stream flows (by reducing the amount of wastewater returned to the stream) with serious consequences for the environment."[7]

To be sure, desalination and other high-tech solutions will help wean the nation of its dangerous overreliance on groundwater. So will more efficient fixes, such as conservation and true-cost pricing. But if the history of water in the United States teaches anything, it is that engineered solutions are rarely as perfect as their boosters boast. Consider President Herbert Hoover's Cooke Commission on Water Resources, which predicted, in 1950, that the nation's rainfall would be doubled by cloud seeding. (The commission also ignored the water needs of fish, wildlife, and recreation and said the federal government should wait a decade before intervening to control water pollution.)[8] Like the dams and the wetlands-drainage and the dredging schemes of the past, scientific solutions usually come with unintended consequences. One of those can be found deep underground, just a few more miles down Kings Highway.

TROUBLE UNDERGROUND

Not far from FEMA City, after Kings Highway passes under Interstate 75, it narrows back to two lanes heading into Florida's interior. Past a few moderately priced new subdivisions, the land begins to give way to cattle ranches once again. Retail development remains a few years away. For now, the commercial strips between the tract homes are filled mostly with businesses related to the home-building industry: contractors, real-estate agents, land surveyors, engineers. And then, around a big curve, the business most important to the rooftop crops: water supply.

At first glance, the Peace River/Manasota Regional Water Supply Authority looks like your everyday water plant: a pair of huge, blue cylinders to store water and to treat it. But what is interesting about this place is not what you can see—it is what you can't. Deep in Florida's limestone aquifer here lies the biggest underground water-storage project in the eastern United States.

Known as Aquifer Storage and Recovery, or ASR for short, the technology was designed to withdraw millions of gallons of water during wet months and to store it deep underground to retrieve during dry months. The Peace River facility, which stretches across 6,000 acres on both sides of Kings Highway, includes a 600-million-gallon reservoir and 21 wells. Each blue well is surrounded on all sides by a chain-link fence, giving the well field the appearance of a roadside zoo full of pipe animals.

When water is plentiful, the authority sucks 36 million gallons a day from the nearby Peace River, a watershed that begins in Central Florida's Green Swamp and flows south 105 miles into the Charlotte Harbor Estuary. It goes first to the reservoir, then to the treatment plant. Twelve million gallons are saved in the reservoir, 18 million flow to the four fast-growing counties served by the water authority. The remaining 6 million gallons are injected via the ASR wells between 700 and 900 feet underground. There, the water sits in huge bubbles in the aquifer, which acts as a natural underground storage tank. During dry months, the ASR wells work in reverse: sucking the water back up again to meet demand.

The U.S. EPA estimates there are 1,695 ASR wells in the nation, most of those in the water-troubled states of California, Nevada, Texas, and Florida. The Peace River water authority was home to the first ASR well in Florida. It was sunk in 1983, not by the government, but by a home-building company called General Development Corporation, which later went bankrupt.

Miami-based GDC started out as a family business in the 1950s and grew into a $500 million a year conglomerate that was famous for the hard sell. In its heyday, it had a 3,000-person sales force that sold 400,000 lots and more than 30,000 homes throughout the state. The company played a huge role in luring northern retirees to South Florida. But by the early 1980s, GDC was in deep trouble, facing steep debts and investigations into allegations that it cheated customers.[9]

Some employees who worked for GDC's water department back then still work at the now-government-owned Peace River facility. In the early 1980s, GDC's utility company had to expand water supply for its fast-growing southwest Florida developments such as North Port. A reservoir was the logical choice. But a brand-new technology called ASR was available for about a tenth the cost. "They did ASR out of desperation," recalls one employee. "The decision wasn't based on hydrologic data at all. It was based solely on cost."

Turns out, hydrologic data is crucial to sinking a well. That is true for all sorts of reasons, not least of which is the concentration of arsenic in some parts of the aquifer. It is not uncommon for arsenic to show up in groundwater in some parts of the country, mostly through erosion of natural deposits, and sometimes from industrial runoff. Some geologic formations, including some in Florida's aquifer, hold much higher concentrations of it than others.

Arsenic, a metal, is the fifty-second-most-abundant element on the planet, averaging 2 parts per billion (ppb) of the earth's crust. It also can be deadly to humans. The French emperor Napoleon Bonaparte likely died of arsenic poisoning; modern scientists confirmed toxic levels of arsenic in his hair, though they disagree about whether he was deliberately poisoned by his enemies or died as a result of gradual exposure.[10]

The EPA classifies arsenic as a human carcinogen. It can cause skin, bladder, lung, kidney, liver, prostate, and other cancers. It is called an "accumulative enabler" because it makes people more likely to become ill from various cancers, diabetes, and high blood pressure. It may cause cardiovascular, pulmonary, immunological, neurological, and endocrine diseases.[11]

The EPA classifies arsenic at a concentration of 60,000 ppb "immediately dangerous to life or health," meaning if you ingest that much, you will die. Ingestion at levels between 300 and 30,000 ppb may cause stomach pain, nausea, vomiting, and diarrhea. Under the Safe Drinking Water Act, the EPA has regulated arsenic in drinking water since 1976. For twenty years, the highest acceptable level was 50 ppb. In 2000, the Clinton administration proposed lowering that to 10 ppb based on new evidence of increased risks of bladder and lung cancer. After reviewing the proposed change, President Bush's first EPA secretary, Christine Todd Whitman, approved the new standard and made it effective January 1, 2006. (When she was governor of New Jersey, Whitman had lowered the acceptable levels of arsenic to 10 ppb in that state.)[12]

When the standard changed, lots of Florida utilities managers found themselves in a bind. Testing by the state's Department of Environmental Protection revealed that numerous ASR wells in Florida contained arsenic-contaminated water greater than 10 ppb. Some wells far exceeded the previous standard of 50 ppb. How could river water with no trace of arsenic become contaminated during underground storage? Scientists soon had an answer. Arsenic is ubiquitous in the part of the aquifer where ASR wells store water; in Central Florida this is known as the Suwannee Limestone layer. The mix of highly oxidized ASR water with the low-oxygen water that exists in the Suwannee Limestone causes arsenic underground to "mobilize."[13] Hydrogeologists are trying to figure out how to stop it. Engineers believe that running the same water through ASR wells several times will alleviate it. But EPA rules prevent them from serving up water from wells that have tested so high. At an ASR conference in Florida in 2005, scientists and utility directors literally spent more time talking about how to finagle exemptions to the drinking-water standards than they did discussing solutions to the arsenic problem.[14]

The problem proved daunting enough that the Peace River authority backed off plans for a major expansion of ASR wells to supply water to its burgeoning population. Instead, the authority is building a 6-billion-gallon reservoir. It will be ten times bigger than the current reservoir, and cost $55 million. Executive Director Patrick Lehman estimates that is about 20 percent more than new ASR wells would have cost. "I think the science will catch up," Lehman says of the arsenic problem.[15]

"Science will prove sound methods to solve the problem," he says. "The question is when."

In other parts of the nation, too, governments have halted ASR plans over arsenic concerns. In 2003, the Wisconsin Department of Natural Resources stopped the City of Green Bay's pilot project to store Lake Michigan water in ASR wells after tests revealed the underground storage hiked arsenic levels.[16] And arsenic is but one wild card in the ASR deck. The Georgia legislature has placed a moratorium on any ASR wells in the Floridan Aquifer, which underlies southeast Georgia and parts of South Carolina and Alabama in addition to Florida, over concerns that stored water could contaminate the deep aquifer water. Other fears include land subsidence or the fracturing of aquifers.

Despite these uncertainties, the federal and Florida governments maintain an astounding level of faith in ASR technology. It is a cornerstone of the Comprehensive Everglades Restoration Plan now underway in South Florida, representing one-fifth of the estimated $10 billion total restoration cost. The plan calls for 333 ASR wells, which would be the largest use of ASR in the world. They would pump as much as 1.7 billion gallons a day of excess surface water and groundwater during wet times and store it in the Floridan Aquifer to recover during seasonal dry periods or longer-term droughts. Experts, such as a National Research Council panel advising the restoration effort, have raised numerous concerns over whether ASR will work in the Everglades. Those include "suitability of proposed ASR source waters, paucity of regional hydrogeologic information, hydraulic fracturing of the aquifer, impacts on existing wells, water-quality concerns, mercury bioaccumulation and others."[17]

Three pilot studies underway in South Florida over the next eight years will help determine whether the Army Corps and South Florida Water Management District go forward with the ambitious ASR plans. "We still don't have the science to understand what happens when you put that much water in the ground," admits Chip Merrian, deputy executive director for water resources at the district. If the pilot projects prove the area unsuitable for all or some of the 333 ASR wells, he says, the agencies will change course. That is the idea behind "adaptive management," the core philosophy of the thirty-year restoration plan. Scientists and engineers can try something, see how well it works, and then modify it.

Just in case the enormous underground storage proposals fall through, water managers are lining up other ways to get water to the masses crowded onto the southeast tip of the state. South Florida's population is expected to double, from 7 million people today to 15 million by 2050. One option is a mammoth desalination plant somewhere on the lower southeast coast that could make up to 30 million gallons of freshwater a day, about the amount consumed each day by the residents of West Palm Beach. Such a plant could be finished by 2015, water managers say, as much as fifteen years ahead of many of the water-storage schemes in the Everglades plan.[18]

Before they go forward with a plan so bold, they are likely to look closely at how a similarly ambitious one has gone all wrong, just across the state in Tampa Bay.

DESALTING THE SEA

The first time Americans looked to the sea for drinking water was in 1861, at Fort Zachary Taylor in Key West, Florida. The Union army ordered a "Marine Aerated Fresh Water Apparatus" from England, patented by a man named Alphonse René le Mire de Normandy of Judd Street in London in 1853. The huge device, shipped across the Atlantic Ocean to New York and then down to the Florida Keys, sucked salt water from the sea, heated it in a giant boiler, and sent it into towering pipes, where it was evaporated and condensed, leaving the salt behind. Fueled by wood and coal, the apparatus provided 7,000 gallons of freshwater a day to the 500 soldiers who were stationed at Fort Zachary Taylor between 1861 and 1865 to protect the nation from enemies who might attack from the Florida Straits.[19]

Today, Fort Zachary Taylor is a national park, where tourists can learn about the Civil and Spanish-American wars (actually, most people head there because it has the most beautiful beach in Key West). Park services specialist Harry Smid maintains historic documents relating to the desalting apparatus. "You would think," Smid muses, "that if the Union Army could de-salt the sea in the 1860s, that we could do it today."

Easier said, the cliché goes, than done. Wresting salt from the sea was so difficult, and so inefficient, that it would be easier to ship freshwater down from Tampa in barges, which is just what Conchs, the nickname for Florida Keys natives, did later in the nineteenth century.

Still, Floridians would not give up on the dream of putting their vast seawater supply to work. One hundred and fifty years after the army brought what is now known as desalination to the United States, Florida set two more milestones in the history of the technology.

First, it would build the largest desalination plant in the Western Hemisphere.

Then, that plant would become the most spectacular failure of desalination technology anywhere in the world.

In 1961, President John F. Kennedy spoke of the great promise of desalination to solve the global freshwater crisis: "If we could ever competitively, at a cheap rate, get fresh water from saltwater," he said, "that would be in the long-range interests of humanity (and) would dwarf any other scientific accomplishments."[20]

"Cheap" would prove problematic. Instead of solving water crises in those parts of the world where people die each day for lack of freshwater, desalination plants tend to be built by the richest nations. There are two primary ways to remove salt from water. One is Alphonse René le Mire de Normandy's method: heating the water and condensing the steam, which is called "distillation." The more modern technique is to filter the water through a membrane, known as "reverse osmosis," or RO. Both take huge amounts of energy, and for that reason, desalination is by far the most expensive of all water-supply options.[21] That made it the purview of wealthy countries in the Middle East. Energy-rich but water-poor nations have built plants throughout the Persian Gulf.[22] Saudi Arabia is the world's largest producer of desalinated water, with more than thirty plants that provide 70 percent of the kingdom's water supply.

Until the late 1990s, only two American cities had built true desalination plants—Key West and Santa Barbara, California. But costs were so high that both cities shuttered the plants soon after opening them. Today, Key West and Santa Barbara maintain them only in case of an emergency.[23]

As RO technology improved, and as costs dropped, more and more cities around the world began to look to the sea for water. RO proved efficient for desalting brackish waters, too, at lower costs than seawater desalination. By 1998, more than 12,000 desalting plants had been built in more than 120 countries.[24] One lesson they took from Saudi Arabia was this: costs could be further reduced by building a desalination plant next to a coastal power plant, which already sucks up and filters millions of gallons of seawater each day—and which offers inexpensive electricity.[25]

All of this made a city called Apollo Beach, home to Tampa Electric Company's Big Bend Power Plant on brackish Tampa Bay, seem like the perfect place to build the largest desalination plant this side of Saudi Arabia. In 1998, the nearly twenty governments that had been warring over the disappearing groundwater of Tampa Bay declared a truce. They joined together to establish a regional water utility called Tampa Bay Water. The utility is the largest water wholesaler in Florida, selling to utilities in Hillsborough, Pinellas, and Pasco counties that in turn sell to 2.5 million people. The idea was that one big utility could eliminate economic competition for water, and focus on regional supply and conservation planning to slash dependence on groundwater in half. The Southwest Florida Water Management District would help the utility fund new supply as long as it met specific goals for cutting groundwater use over time: from 192 million gallons a day to the eventual low of 90 million gallons a day by 2008.

In 1999, Tampa Bay Water approved plans to build a $110 million desalination plant at Apollo Beach that would churn out 25 million gallons of freshwater a day. Desalination boosters around the world were excited about the plant because of the apparent breakthrough in price it would achieve. Relatively low salinity in the bay, cheap energy, and sharing infrastructure with the Big Bend Power Plant all helped lower costs. The desalination plant would be privately owned and operated by a consortium including Poseidon Resources and Stone and Webster Company. The team committed to begin operating the plant in 2002, and to deliver desalinated water at an unprecedented wholesale cost of just under $2 per thousand gallons. At the time, large-scale seawater desalination cost between $4 and $6 under "optimum" operating conditions. It could cost as much as $10 per thousand gallons if the plant did not operate efficiently.[26] By comparison, it costs about $1 per thousand gallons to pump groundwater and deliver it wholesale.[27]

Looking back, Ken Herd, Tampa Bay Water director of operations, says his board never could have anticipated the bad luck that would dog the project. In 2000, Boston-based Stone and Webster, one of the world's largest and most respected engineering firms, declared bankruptcy. Its partner, Poseidon, then hired New Jersey–based energy giant Covanta to complete the plant. A year later, Covanta filed for bankruptcy after the energy crisis in California crippled its cash flow.[28]

Tampa Bay Water decided to take over ownership of the plant. It had access to cheaper financing than the private companies. Moreover, the most significant hurdle, environmental permitting, had been jumped.

The utility ousted Poseidon but stuck with Covanta after the bankruptcy so the company could finish the plant it started. A spin-off company called Covanta Tampa Construction completed the plant in the spring of 2003. On the day it produced its first 3 million gallons, Tampa Bay Water officials toasted with plastic cups. But the celebration was premature. When the plant began to produce near capacity, its expensive membranes started to clog. Covanta officials blamed invasive Asian green mussels that had been introduced into Tampa Bay in ship ballast water and grew on the intake pipes at the power plant, which provides the water to the desal plant. Tampa Bay Water officials believed the real culprit was Covanta's pretreatment system. Later that year, Covanta Tampa Construction went bankrupt before Tampa Bay Water could fire it.[29]

As this book went to press in 2006, the Tampa Bay Seawater Desalination Plant was still shuttered. Tampa Bay Water hired yet another impressive team of experts to get it up and running. American Water-Pridesa LLC is a joint venture between American Water Services and Pridesa S.A. of Spain. The team bid $29 million to fix the plant's pretreatment system, which the utility says is not rigorous enough to filter particles from seawater, causing the cartridge filters and desalting membranes to clog too quickly. Overall costs now have surpassed $148 million. The latest price estimates per thousand gallons of water, once an impressive $2, had climbed to $2.54. That was "still the lowest cost among operating desal plants of similar size throughout the world," according to Tampa Bay Water spokeswoman Michelle Robinson.

Herd and other Tampa Bay Water officials had faith that the plant would be up and running by 2007. If it is, the most interesting part of the story will still be this: over the years that the plant's troubles dragged on, the region managed to reduce its groundwater pumping from 192 million gallons a day to 121 million gallons a day despite population growth and *without* the desalted water the officials insisted they needed to meet that goal. Instead, a new, 15-billion-gallon reservoir and a 66-million-gallon surface-water treatment plant, combined with aggressive conservation initiatives, had slashed groundwater use in the region by more than a third without a drop of desalted water.

Herd says Tampa Bay just got lucky. Plenty of rain fell in the years the plant was shuttered. "The wisest thing we've done is diversify our water supply," he says. "The goal is to have a diverse, drought-resistant water portfolio, just like having a diverse investment portfolio. We will need desalination in times of drought."[30]

Still, it is a hugely expensive technology in places such as Florida that have not met efficiencies in water pricing or conservation. Even in oil-rich Saudi Arabia, Crown Prince Abdullah recently cited the high energy costs of desalting more than 570 million gallons of water a day when he urged residents to do a better job of conserving.[31]

Nature, itself, is a desalination plant of sorts. Heat from the sun evaporates water from the ocean's surface. The water vapor eventually comes into contact with cooler air, where it recondenses, then falls to earth as rain and dew. With its 1,400 miles of coastline and 55 inches of rainfall each year, Florida is a huge benefactor of this natural process. But even at that macro scale, desalination cannot slake the state's growing thirst.

Are desalination and other technologies really the best solutions? Or do they aggravate water-supply problems by enabling more and more people to squeeze into fragile places that do not have the water to support them? If desalination could save the 9,500 children who die each day because of lack of good water, it could well, in President Kennedy's words, "dwarf any other scientific accomplishments." If it serves only to wedge more people into Florida, California, and Texas—the three fastest-growing and heaviest-water-using states are also the three biggest players in desalination—its promise is not so noble.

12

Redemption & the River of Grass

TWENTY-FIVE YEARS AGO, a bankrupt commercial photographer from California did what so many other down-on-their-luck Americans have done when they needed to make a fresh start. He packed all his stuff into a truck, loaded up his family, and drove down to Florida.

Clyde Butcher had earned his living mass-producing color lithographs of natural scenes for American living rooms. Sold through J.C. Penney, Sears, and other department store chains whose sales records told him what photos would sell—seagulls yes, mountains no—some of Butcher's images would sell 250,000 times over. But the man so good at making beautiful pictures was not so at finances. Things as mundane as accounts receivable brought him down, even as his photographs sold by the thousands.[1]

A cross-country road trip with their nine-year-old daughter and seven-year-old son was not a bad thing for Clyde and Niki Butcher. They were hippyish nature lovers who had lived in a tent trailer in California's state parks when the kids were babies, aboard a twenty-six-foot sailboat when the children were a little older and Butcher was making money. The couple wanted to find a place where they could sail with their son and daughter, and where Butcher could take the sorts of sunset-and-seagull shots that make Americans reach for their wallets. South Florida had the seagulls and the sunsets. And it was a sailor's paradise, with quiet bays and gentle waters compared to California's violent waves.[2]

The Butcher family's early years in Florida were not smooth sailing. As everyone who runs to Florida figures out eventually, you cannot leave your troubles at the state line. When they first arrived, in 1980, creditors from California tracked them to Fort Lauderdale. Repo men hauled off their car, van, and motor home.

But over time, with the help of a friend, they started selling Butcher's photographs once again. Back then, his photos were known for little clocks mounted in the corners. He and Niki cranked them out in a rented garage in Lauderdale. Eventually, the Butchers turned to the art-show circuit, selling the clock pictures at weekend festivals up and down the eastern United States. The family began to thrive again. They settled in Fort Myers, on Florida's southwest coast.

Butcher did not get much satisfaction from his work. He knew his pictures were sappy and commercial. He also was not enamored with Florida—except, that is, for the sailing. "Whatever there once was of value," he used to say, "had been turned by the plow or paved over."[3]

But Butcher was not a big complainer. He and Niki had everything they moved to the Sunshine State for: the sunset-that-sells scenery, the sailing, the closeness (sometimes too much) with children Jackie and Ted, now teenagers.

They would have all that until Father's Day in 1986, also Clyde and Niki's twenty-third wedding anniversary. The knock on their door came at midnight. Their son, the sweet and introspective Ted, was dead at seventeen. He had been killed in a car crash, by a drunk driver.

The story of Florida's vanishing water is not just one of senseless waste. It is also a story of redemption. Redemption can take many forms. For some, it is as easy as a baptism in the Suwannee River, an instant washing away of sins. For others, it is a lifelong struggle. Some find redemption in a child, others in a church. Still others find it in nature.

Clyde Butcher found it in the Florida Everglades.

In the weeks following Ted's death, Butcher began to wander Big Cypress National Preserve, taking solace in its beauty and solitude. He found a Florida he never knew existed: waves of marsh grass as far as he could see; cypress stands as stately as California's Avenue of the Giants; the huge, primeval tree islands and the tiny, delicate orchids; all under a sky as wide as an African savannah.

After enduring funerals for first his son and then his father just two weeks later, Butcher gathered up his life's work—color negatives and prints worth more than $100,000—and tossed them. He bought an eight-by-ten-inch box camera that he could not afford. And he started shooting the Big Cypress in black and white. It was the way he used to shoot California's mountains when he was young, before J.C. Penney told him that mountains do not sell.

Butcher's new, oversized photos of Florida were haunting and primal, not a seagull or a sunset in sight. The first time he showed them in public, at an art festival near Orlando, people swarmed his booth. "Is this Africa?" they would ask. "Is this the Amazon?" And he would answer, "No, it's just out in the Everglades."

"People had never seen pictures that allowed them to feel the Everglades," Butcher says. And the pictures sold. Better than his sappy, clock-affixed pictures ever had.

Today, Butcher is one of the best-known photographers in the United States. His work hangs in the homes of movie stars and U.S. senators. When *Popular Photography* recently asked, "Who is the next Ansel Adams?" the magazine named Butcher one of the nation's four "keepers of the large-format flame."[4] His business, now run by his daughter Jackie, has grown as large as his photos, as big as his three-hundred-pound frame.

But for Butcher, photography is no longer business. As he trudged deeper and deeper into the Everglades, he came to see that the swamp was dying—a victim, like his son, of human carelessness. In this case, Butcher might be able to do something about it.

In the 1870s, landscape artist Thomas Moran's paintings of Yellowstone cinched the campaign to make it a national park. In 1938, Ansel Adams's book of wilderness photographs, *Sierra Nevada: The John Muir Trail,* helped convince Franklin Roosevelt's administration that national parks should support wilderness, not resorts.[5]

Butcher's Everglades photos became popular just as the notion of "Save Our Everglades," Governor Graham's restoration project, was taking hold in Tallahassee and Washington, D.C. Butcher became a champion of Everglades restoration, a particularly effective one because he

The Arthur R. Marshall Loxahatchee National Wildlife Refuge, the last northernmost portion of the unique Everglades, was named for the crusading biologist who convinced Florida's politicians that the great swamp could be restored. (Photograph courtesy of Clyde Butcher.)

could show someone who would never step foot in the Glades why they were worth saving.

As he sold thousands of prints, Butcher also gave his photos away to any environmental organization or government agency that needed them for education or promotion. His ancient-looking forests, rivers, and prairies helped change the way Americans saw the Everglades. The photos were too beautiful to reveal the destruction of the ecosystem—even his photos of cattails, which signal pollution. But they made people fall in love with a swamp that previous generations had disdained and destroyed.

It was, for Butcher, redemption. "Before the death of our son, the images were products," he says. "After Ted's death, they became art that could educate people about the loss of the world around them. When something like that happens, you can either become positive or negative.

"If you become negative, you've wasted a soul."

Today in the Everglades, in the ambitious experiment to try to undo 150 years of draining, channeling, polluting, and other abuse, the Florida and

U.S. governments are flooding former cattle lands, blowing up dams, and building huge reservoirs, all part of the largest public-works project in the history of the world. The Comprehensive Everglades Restoration Plan was, at last count, slated to cost $10.5 billion and to take at least thirty-five years.

Fixing the Everglades was a chance for redemption. It was a chance for South Florida's local government officials to show they could stand up to political pressure and stick to the development maps they had drawn. It was a chance for Jeb Bush to prove his intentions to protect Florida's environment were equal to those to find water for growth. It was a chance for the U.S. Army Corps of Engineers to demonstrate the agency that had done so much harm to America's waterways was capable of a turnaround: "We are putting the best and the brightest minds and everything we've got into getting this right," said Colonel Robert Carpenter, Corps commander in Florida.[6]

In fact, restoring the Florida Everglades was a chance at redemption for all America. If the country could fix its worst water screwup in the eastern United States, it could fix the others, too. If Florida could get thousands of acres of wetlands back, Louisiana could get thousands of acres of wetlands back. If parts of the Everglades could look like they did a century ago, so could parts of the Chesapeake Bay. Some Everglades scientists had headed to the great Pantanal, ten times bigger than the Glades at 36.5 million acres, to help people in Paraguay, Bolivia, and Brazil avoid America's ecological mistakes.

If, with a combination of conservation and sound water-supply projects, South Florida could come up with its own water for growth, people in North Florida could put up their shotguns. Great Lakes environmentalists could take down those "back off suckers" billboards. In fact, if the Corps' engineers could turn the C-38 canal back into the meandering Kissimmee River, who is to say that engineers cannot someday restore the Colorado River to its free-flowing state, a dream of the board of the Sierra Club?

Those, of course, are a lot of "ifs." At this writing, the Comprehensive Everglades Restoration Plan, signed into law with so much bipartisan good will and hoopla on the very same day in 2000 that the U.S. Supreme Court was deciding *Bush v. Gore,* seemed to be stuck in the muck.

One problem was that Congress had not anted up its half of the restoration money. Congress was supposed to fund the Comprehensive Everglades Restoration Plan regularly in its biennial Water Resources Development Act. But the nation's water bill had not passed for six years, from 2000 to 2006. Versions of the WRDA bill passed the House and Senate in fall 2006, and it was pending in conference committees as this book went to press. Water projects across the country sank as lawmakers argued for several years over Army Corps reform. And then, they were presented a far more intractable problem: how to restore coastal Louisiana and Mississippi in the wake of Hurricane Katrina.

But the biggest threat to Everglades restoration was the one that led to the marsh's demise in the first place: the force of growth and development as that net total 1,060 people continued to move to Florida every single day. In South Florida, the force was so powerful that acreage targeted for environmental restoration was being bought up and paved over faster than the state could bid on it.

By the end of 2005, the state of Florida had purchased half the lands required for restoring the Everglades, about 200,000 acres. Development and price pressures were rapidly putting the other 200,000 out of reach. Scientists watched as lands identified for Comprehensive Everglades Restoration Plan projects turned to rooftops instead: Two hundred acres for a water-quality project in the Biscayne Bay Coastal Wetlands became suburban housing. Nearly two thousand acres for a wetland restoration project near the Indian River Lagoon became ranchettes.[7]

Meanwhile, local government officials in South Florida were back to pushing their urban growth boundaries on behalf of the powerful homebuilding industry. In one of numerous examples, Miami-Dade County asked state regulators to consider more than a dozen exceptions to allow the spread of suburbs outside its development line, including a 6,000-home development in a low-lying area between Biscayne and Everglades national parks. This is where the Army Corps gave a company called Atlantic Civil a permit to fill more than 500 acres of wetlands for agriculture. Three years later, Atlantic Civil instead filed plans for a brand-new town of 18,000 residents, including the 6,000 homes to be built by Lennar Corporation, one of the largest public companies in Florida.

It is more than a tad hypocritical, of course, to ask taxpayers spread across the nation for money to restore the Everglades when the local governments who live next door are not doing their part to protect the marsh. It is sort of like Florida demanding Georgia reduce its water withdrawals to set aside more of the Chattahoochee for natural systems downstream: Florida slurps up 8.1 billion gallons of water a day to Georgia's 6.5 billion.[8]

With the blessings of Governor Bush, state regulators advised against every one of Miami-Dade's proposed exceptions, saying, in part, that the county did not have enough water to supply the new growth. It was the first sign that Florida's new water law was having an intended effect, linking development to the availability of water.

That was just one in an unusual stretch of triumphs for Florida's environmental defenders. In late 2005, a federal judge halted the Army Corps' permit that gave the St. Joe Company developing Florida's panhandle broad latitude to wipe out 2,000 acres of wetlands. It was the third time in a year that federal judges overturned permits to destroy Florida wetlands for development and mining.[9]

Then the Army Corps—despite political pressure from U.S. Senator Bill Nelson, a Florida Democrat, and former U.S. representative Porter Goss of Sanibel, a Republican who later led the CIA—rejected a controversial golf course development near the Everglades that would have destroyed thousands of acres of wetlands. The 800-home Mirasol development in southwest Florida had won approval from local officials, state environmental regulators, and the South Florida Water Management District. Mirasol's houses and two eighteen-hole golf courses would have been built on 1,766 acres near Bonita Springs, 1,500 acres of which were wetlands. Mirasol would have wiped out 587 of those acres. Worse, a three-mile-long ditch to funnel storm water around its houses would have drained an additional 2,000 acres nearby, the EPA reported.[10]

Nathaniel Reed was the grand guard of Florida's environmental movement since the death of Marjory Stoneman Douglas in 1998 at age 108. Reed, former governor Kirk's dollar-a-year environmental adviser, had battled with the Army Corps for half his life. Yet he lauded Colonel Carpenter for bringing "vigor" to the Florida Corps, as well as "passionate, young, environmental-minded" engineers and scientists who were working to restore the Everglades.[11]

The long century of abuses that destroyed Florida's Everglades, and the complicated efforts to restore the marsh, had short, simple lessons. Water is best stored in its natural systems: wetlands, streams, and rivers. Huge waterworks are not only the most expensive and environmentally damaging solutions to water woes but they seem to cause more problems than they solve. Finally, it is easier and cheaper to prevent ecological damage than to reverse it.

Look at the Kissimmee River. Taxpayers spent $35 million to turn the river into a canal, hauling in 3 million truckloads of dirt and building five dams. Taxpayers now are spending more than ten times that, $500 million, to restore the Kissimmee.

One lesson Florida has not learned, and maybe it never will, is that increased growth and economic prosperity do not have to equal increased water consumption. Water use in the United States stopped rising in the 1980s, yet population as well as gross domestic product have grown steadily ever since.[12] Think tanks such as the Rocky Mountain Institute are brainstorming a "soft path" for water that emphasizes greater efficiency, more precise management systems to avoid waste, and better matching of water sources to their uses. For example, supplying drinking-quality water for consuming and cooking but not for irrigation or toilet flushing. Half of all water treated to meet federal EPA standards for drinking goes down the toilet.[13]

Since 1990, water use in Southern California has dropped by 16 percent, even as the population has increased by nearly the same figure. In Seattle, total water use has remained constant since 1975, even though population has increased by 30 percent.[14]

Water use in the greater Boston area hit a fifty-year low in 2004, following an aggressive conservation program begun in the late 1980s that has indefinitely postponed construction of a diversion from the Connecticut River and saved residents more than $500 million in capital expenditures alone.[15]

"The fastest, cheapest, and most environmentally acceptable way to address water conflicts will in most cases not be an increase in supply, but improvements in efficiency to reduce waste and increase water supply reliability," says Peter Gleick. "Realizing these savings will be faster, cheaper and more politically acceptable than any new supply option proposed, including new dams, desalination plants or long-distance aqueducts."[16]

In October 2005, six weeks after Hurricane Katrina plowed into Louisiana and Mississippi and drowned more than one thousand souls, a hurricane called Wilma whirled into South Florida. Unlike the people of New Orleans and their neighbors, the vast majority of Florida's suburbanites were spared calamity. A record 6 million people lost power, but most had a roof over their heads, running water, even open grocery stores. You would not know it, though, from the ensuing pandemonium. Despite all the standard warnings to gas up the car and have three days' water and groceries on hand, many South Floridians, just twenty-four hours after the storm, queued up in long lines to wait for water and ice. They idled at gas stations for hours to fill up their SUVs. "I think they've forgotten about us," cried a woman who waited two hours in suburban Kendall to buy gas for her generators that powered her two refrigerators.[17]

"We need food! We need water! We need checks!" said another woman two days after the storm hit, as she waited in line for three hours at Salvation Army to eat a lukewarm chicken breast. Her home, she said, had not been damaged.[18]

TV reporters interviewed middle-class guys standing in long lines for free water in a Wal-Mart parking lot, even though gallons were stocked and on sale in the store. Cameras captured others munching Burger King as they waited in line for free food. Florida, as seemed more and more common, became a punch line for the late-night television comedians, a punching bag for the daytime talk-radio conservatives.

Even the liberals had to nod their heads when Governor Bush observed: "People had ample time to prepare, and it isn't that hard to get 72 hours of food and water, to do the simple things that we ask people to do."[19]

Waiting by their cars in a Miami gas station line, a group of Haitian men chatted in Creole about how losing power would not have crippled them in Haiti, the poorest country in the Western Hemisphere. "We've been Americanized too much," said Jaques Edward. "If we were living like Haitians, we wouldn't be in trouble."[20]

Indeed, these helpless, impatient Floridians had nothing in common anymore with their rugged predecessors who homesteaded the state a century and a half before. Those scrappy settlers faced down alligators, swarming clouds of mosquitoes, and worse as they worked under the Florida sun to tame a swampy paradise in the days before air-conditioning.

Floridians, it turns out, endangered more than the panthers and the plume birds when they filled in wetlands, paved over swamplands, drained ponds and lakes, bulldozed wet jungles, and tapped out groundwater. They did not just get rid of the baby-fish nurseries and the bird rookeries. They did not only destroy their best tools for flood control and erosion. They did not simply wipe out their drinking-water supply.

Draining the water drained Florida's cultural identity and, it is no exaggeration to say, the spirit of its people.

Like many other Americans, most Floridians once had a daily connection with the water and the land even if they worked desk jobs. Driving across town, they would see a shimmering lake or the sea, tall pines and mossy oaks, a swooping hawk, a grassy marsh or two. By the twenty-first century, though, local government officials in most coastal counties had allowed so many tall condominiums on the beaches it was hard to remember Florida was a peninsula. Home builders had flattened so many native trees and built so many soulless subdivisions that pulling off Interstate 75 in Tampa felt no different from pulling off in Macon, Chattanooga, or Lexington.

Even the islands did not feel like islands. If you were blindfolded and driven to the middle of Marco Island, just a block from the breathtaking Gulf of Mexico, when you opened your eyes you'd have no clue you were anywhere near the sea. You'd see a tall wall of hotels and condos; that's all.

President Theodore Roosevelt always feared America's destruction of its natural places would bring about a national malaise. He believed, like Frederick Jackson Turner, that Americans' proximity to wilderness built a national character of ruggedness, resilience, and ingenuity.[21] His study of U.S. history, along with his personal experiences hunting, fishing, and camping, convinced him that living with nature promoted "that vigorous manliness for the lack of which in a nation, as in an individual, the possession of no other qualities can possibly atone."[22]

Roosevelt saw the importance of Americans keeping close to their environment as no less than a matter of national security. The modern American, he wrote in 1899, was in real danger of becoming an "over-civilized man, who has lost the great fighting, masterful virtues."[23]

In his book *Last Child in the Woods,* journalist and child advocate Richard Louv weaves scientific research with the words of children to show how the absence of nature is aggravating some of the most disturbing childhood trends: the rise in obesity, attention-deficit disorders, and depression. Louv came up with the phrase "nature-deficit disorder" to

Water, water everywhere but not a drop to drink. Miami is surrounded by water but faces freshwater shortages due to overpumping of groundwater and population growth. (Photograph by Art Seitz/Silver Image.)

Fakahatchee Strand, adjacent to Big Cypress National Preserve, was plotted for a subdivision before the state and federal governments purchased it in the 1970s. (Photograph courtesy of Clyde Butcher.)

describe "the human costs of alienation from nature, among them: diminished use of the senses, attention difficulties and higher rates of physical and emotional illnesses." In just one of hundreds of examples, Louv cites new research that suggests exposure to nature may reduce the symptoms of attention deficit hyperactivity disorder (ADHD), and that it can improve all children's cognitive abilities and resistance to negative stress and depression.[24]

Such results make sense to anyone who has seen how an hour hunting shark's teeth in a creek can calm, and conk out, a nap-fighting four-year-old. But too many kids have been cut off from creeks, wetlands, and rivers (not to mention treehouses and vegetable gardens, if they live in one of the many deed-restricted communities that prohibit those so-called eyesores). In Miami, a metro area surrounded by water, kids are so removed from nature that the Miami River Commission takes schoolchildren on field trips to the waterway that they would never otherwise see, even though it runs through the heart of their city. For adults there, just getting to work on the massive concrete highway mazes can be so stressful and unsightly, even on days when everything is running as it should, it is no wonder the two-fridge soccer mom told a reporter that Hurricane Wilma's power outage had caused her to have a panic attack.[25]

From coast to coast, Americans have lost touch with their water. In California, the San Joaquin River no longer makes it to San Francisco Bay. In Florida, children no longer swim in Crystal Springs because of a water-bottling operation. This loss has consequences far beyond stress or aesthetics. It is a matter, as Theodore Roosevelt foretold, of our national heartiness, our fitness and vitality as a society. In the wake of a national disaster or a terrorist attack, who would you rather be with? The guy in the Wal-Mart water line? Or the Girl Scout who, having proved her knowledge of where local streams drain as well as how wastewater is treated, had earned her Water Drop Patch?[26]

Ultimately, the choices we make about water will help determine how we fare as states, as a nation, as humans.

We can go on wasting copious amounts of water, using treated drinking water to quench suburban lawns, or we can appreciate its worth. We can keep giving water away, for free, to anyone with a business plan, or put a price on it to make sure water is protected and valued.

We can continue to bend wetlands and growth laws, or we can demand their consistent enforcement. We can spend more tax dollars on enormous water-diversion and other technological schemes that may be risky, or we can spend them on water-conservation, land-preservation, and restoration projects.

We can watch our children repeat the mistakes of America's water history: in the East, draining water and giving it away to all comers; in the West, damming it up and doling it out until there is not enough for people or nature.

Or we can teach them how lucky they are to have water—for drinking, for bathing, or simply for the sheer joy of plunging into an icy, clear-blue spring on a hot summer day.

Acknowledgments

BOOK AUTHORS seem to thank their spouses last. But this one thanks hers first. It would be impossible for a mother of two young children who has a full-time job to write a book, unless the working mother is married to the likes of Aaron Hoover. Aaron cheerfully engaged our son, Will, who was one year old when I completed the master's thesis that led to this work, and our daughter, Drew, who was one when I turned in the manuscript, for hundreds of hours so that I could complete *Mirage.* Many of those hours were snowy and gray: Aaron took an academic year off his own career track to care for our family in Ann Arbor, Michigan, while I conducted research and interviews during a journalism fellowship.

Aaron is also my first, last, and most trusted editor. As a science writer credentialed in both the classics and fisheries, his brain is full of fascinating facts. It was Aaron who told me Florida's springs were a source of inspiration for one of the most famous poems in the English language, Samuel Taylor Coleridge's "Kubla Khan." Most tales of water shortage in a land of plenty quote Coleridge's "Rime of the Ancient Mariner": "Water, water, every where, Nor any drop to drink." Thanks to Aaron, I get to quote "Kubla Khan" instead. Also for making sure Will and Drew had fun and square meals while their mom was hunched over a laptop, I thank my aunt Mindy Blum, my father-in-law, Paul Hoover, and my mother-in-law, Jane Toby, as adept an editor as a "Nonna."

Next I thank the two groups of professionals with the greatest value relative to appreciation in American culture: newspaper reporters and librarians. *Mirage* was much enriched by librarians and their treasures, from historic maps to hand-edited political speeches, at the University of Florida, Florida State University, the Florida State Archives, the University of Michigan, and the University of North Carolina. And, while I score some scoops in these pages, I also rely on daily newspaper reporting by fellow journalists. Floridians are particularly lucky to have Craig Pittman at the *St. Petersburg Times,* Robert King at the *Palm Beach Post,* and Bruce Ritchie at the *Tallahassee Democrat,* all longtime environmental reporters who serve as watchdogs for Florida's land and water. Thanks to Ritchie, citizens know that our water-quality laws do not go far enough to protect springs. Thanks to King, we know when elected officials renege on promises to restore the Everglades. Thanks to Pittman and Matthew Waite's powerful series, "Vanishing Wetlands," regulators and judges in Florida finally began standing up to extraordinary political pressure to stop the rampant draining and filling of the state's last wetlands. *Mirage* is meant to build upon the work of the journalists, historians, scientists, and others named in the bibliography and endnotes. I am grateful to each one of them.

Another three men who know and care for Florida are those who made up my thesis committee at the University of Florida, where I earned a master's degree in environmental history in 2003. I thank historian David R. Colburn for convincing me that population growth is the single most important story in Florida, and for giving me so much of his time even while he was university provost. I am grateful to Mark T. Brown, an environmental engineer with a passion for Florida history, so aptly nicknamed the Swamp Doctor. And I thank Julian M. Pleasants, who oversees a treasure chest of modern Florida history at UF's Samuel Proctor Oral History Program.

At another university 1,045 miles to the north, I am forever grateful for the University of Michigan's Knight-Wallace Fellows program and its director, Charles Eisendrath, the best-dressed man in the Great Lakes region. In spring 2004, I told Eisendrath and the Knight-Wallace selection committee that I had a book idea I thought was important but I could never pull it off without time away from work, and away from Florida, to research, to write, to think. The following year, the Knight-Wallace Fellowship gave me all that and much more. The program sends fellows to Istanbul, where I got to run my hands along ancient cisterns and aqueducts and to question leaders about Turkey's tense water relations with Syria and Iraq and its grand dam-building plans. Knight-Wallace fellows also travel to Buenos Aires, where I interviewed priests and presidential candidates on water privatization. My professors at the University of Michigan also deserve a great deal of credit for this book: Jonathan Bulkley taught me water policy; novelist Nicholas Delbanco worked with me on narrative writing; evolutionary biologist Bobbi Low introduced me to the Anasazi and other lost cultures whose demise may have had to do with resource depletion.

I also thank the many colleagues, water professionals, friends and relatives, and even a few perfect strangers, who read drafts of this work. I am especially grateful to water-law expert Richard Hamann for his lawyerly catches. Likewise to the eminent economist Sanford V. Berg for his willingness to explain water pricing and economics to a journalist whose description of Adam Smith as "the invisible hand guy" made him cringe. I thank my friend Christine Kirchhoff, an environmental and water resources engineer, for her scrutiny of my science. I thank fellow first-time author Jason Tanz for his comments on early drafts, which went along these lines: "Um, I think this is how you're supposed to write a book." I thank Jim Owens for his careful and caring read and Clive Wynne for catching my Americanisms and more. (What are wingtips? he wanted to know.) I thank my stepfather, Dr. Joe Garrison, whose questions made *Mirage* much better.

Also for their wise feedback and other support I thank Karen Arnold, Denise Cante, Stephanie Milch, Bobbe Needham, Claude Owens, Mary Stone, and my mother, Gerry Garrison. My father, Rusty Barnett, passed on his love of Florida's land, water, and history. I thank my first-ever editor, Jacki Levine, and my most recent one, Mary Erwin at the University of Michigan Press. I am particularly grateful to Erwin and her colleagues at the University of Michigan Press who read my book proposal and understood why Americans would care about disappearing water and wetlands in other parts of their country—before Hurricane Katrina ever hit. Also at the Press, I greatly appreciate the crack copyediting of Betsy Hovey and the design by Jillian Downey.

Finally, I am indebted to my employer, *Florida Trend* magazine, to my *Trend* colleagues, and especially to Executive Editor Mark R. Howard. I accepted my *Trend* position in 1998, figuring I would stay a year or two while I waited for the right newspaper job to open up. My assignments turned out to be so journalistically satisfying, my workplace so humane, that I have never left. In these times of corporate, contracting media, it is no less than remarkable that Florida's business magazine hits so hard on environmental and other issues. Recognizing Florida's land and water are its stock in trade, Howard has given me considerable time and resources for stories about the Everglades, growth management, wetlands banking, water privatization, the bottled-water industry, conservation easements, coastal water quality, and water supply. Each of those stories helped shape this book, as did hundreds of interviews with business owners, citizens, developers, environmentalists, lawyers, regulators, and others I have conducted on behalf of *Trend* over the past decade. I thank them all.

Of course, the opinions and conclusions drawn in these pages are not those of my employer or my editors, my sources, my professors, or anyone else thanked here. They are all my own. And so are any mistakes.

Notes

PROLOGUE

1. Author interview with David and Vivian Atteberry.

2. "Sinkholes, West-Central Florida: A Link between Surface Water and Ground Water," excerpt from U.S. Geological Survey Circular 1182, 1999, 121, 132.

3. Mike Vogel, "Good Migrations," *Florida Trend,* April 2006, 26–31.

4. U.S. Bureau of the Census, http://www.census.gov.

5. Juliana Barbassa, "Future Uncertain for River That Feeds a Nation," Associated Press, October 2, 2005.

6. "Drought Settles In, Lake Shrinks and West's Worries Grow," *New York Times,* May 2, 2004.

7. "Drought Moves into Southeastern U.S." United Press International, April 22, 2004.

8. Author interview with Don Wilhite.

9. Patrik Jonsson, "In Central North Carolina, There's Little Water Anywhere," *Christian Science Monitor,* November 23, 2005.

10. "The Value of Water: Concepts, Estimates and Applications for Water Managers," American Water Works Association Research Foundation, 2005, 189–90.

11. "New Jersey Coastal Plain Aquifer," U.S. Environmental Protection Agency, http://www.epa.gov/Region2/water/aquifer/coast/coastpln.htm.

12. "The Value of Water," 197.

13. "Freshwater Supply," U.S. General Accounting Office Report No. 03–514, July 2003, 60.

14. Ibid., 69.

15. Ibid., 8.

16. "Florida, California and Texas to Dominate Future Population Growth," U.S. Census Bureau Press Release, April 21, 2005. An attached chart shows Florida's population as 17,509,827 in 2005 and 21,204,132 in 2015, a 21.1 percent increase.

17. Bruce Henderson, "Who Gets the Water? The Carolinas Face New Limits as Growth Outpaces Supply," *Charlotte Observer,* December 29, 2002.

CHAPTER 1

1. U.S. Bureau of the Census.

2. John T. Foster Jr. and Sarah Whitmer Foster, *Beechers, Stowes, and Yankee Strangers: The Transformation of Florida* (Gainesville: University Press of Florida, 1999), 89.

3. Nelson Manfred Blake, *Land into Water—Water into Land: A History of Water Management in Florida* (Tallahassee: University Press of Florida, 1980), 83.

4. Marjory Stoneman Douglas, *Florida: The Long Frontier* (New York: Harper and Row, 1967), 159.

5. Raymond A. Mohl and George E. Pozzetta, "From Migration to Multiculturalism: A History of Florida Immigration," in *The New History of Florida,* ed. Michael Gannon (Gainesville: University Press of Florida, 1996), 391.

6. Francis Harper, ed., *The Travels of William Bartram: Naturalist's Edition* (New Haven: Yale University Press, 1958), 69.

7. Roderick Frazier Nash, *Wilderness and the American Mind* (New Haven: Yale University Press, 2001), 40. Nash's is the classic study of American attitudes toward nature, from our yearning to tame it in the early years of the Republic to our yearning to reconnect with it once it was gone.

8. Quoted in ibid., 41.

9. Quoted in Blake, *Land into Water,* 15.

10. "The State of Florida," map, U.S. Bureau of Topographical Engineers, 1846.

11. Stuart B. McIver, *Death in the Everglades: The Murder of Guy Bradley, America's First Martyr to Environmentalism* (Gainesville: University Press of Florida, 2003), 1.

12. David McCally, *The Everglades: An Environmental History* (Gainesville: University Press of Florida, 1999), 61.

13. Quoted in Blake, *Land into Water,* 34.

14. Ibid., 35.

15. Quoted in Nash, *Wilderness and the American Mind,* 23.

16. Blake, *Land into Water,* 35.

17. Mark Derr, *Some Kind of Paradise: A Chronicle of the Man and the Land in Florida* (Gainesville: University Press of Florida, 1998), 86.

18. Sidney F. Ausbacher and Joe Knetsch, "Negotiating the Maze: Tracing Historical Title Claims in Spanish Land Grants and Swamp and Overflowed Lands Act," *Florida State University Journal of Land Use and Environmental Law,* Spring 2002, 355–57.

19. Derr, *Some Kind of Paradise,* 87.

20. Ibid.

21. Michael Gannon, *Florida: A Short History* (Gainesville: University Press of Florida, 1993), 53.

22. Derr, *Some Kind of Paradise,* 89.

23. John M. DeGrove, *Land Growth and Politics* (Chicago: American Planning Association's Planners Press, 1984), 99.

24. Blake, *Land into Water,* 75.

25. Derr, *Some Kind of Paradise,* 87.

26. Ausbacher and Knetsch, "Negotiating the Maze," 357.

27. Gannon, *Florida,* 53.

28. Sidney Lanier, *Florida: Its Scenery, Climate and History* (Philadelphia: J. B. Lippincott and Co., 1876), 16.

29. Foster and Foster, *Beechers, Stowes, and Yankee Strangers,* 88–89.

30. Blake, *Land into Water,* 64.

31. Ibid., 83.

32. Ibid., 92.

33. Samuel Proctor, *Napoleon Bonaparte Broward: Florida's Fighting Democrat* (Gainesville: University of Florida Press, 1950). Quoting the *Volusia County Record,* 241.

34. Gannon, *Florida,* 77.

35. Blake, *Land into Water,* 133–34.

36. Derr, *Some Kind of Paradise,* 55.

37. Charles Torrey Simpson, *Out of Doors in Florida: The Adventures of a Naturalist, Together with Essays of the Wild Life and Geology of the State* (Miami: E. B. Douglas, 1923), 137.

38. Blake, *Land into Water,* 134.

39. Ibid., 140.

40. DeGrove, *Land, Growth and Politics,* 101.

41. Gannon, *Florida,* 85.

42. Ibid., 92.

43. Gary R. Mormino, "World War II," in Gannon, *The New History of Florida,* 323.

44. David R. Colburn and Lance deHaven-Smith, *Florida's Megatrends: Critical Issues in Florida* (Gainesville: University Press of Florida, 2002), 35.

45. Ibid., 38–39.

46. Ibid., 40.

47. Ibid., 41.

48. McCally, *The Everglades,* 76–78.

49. Blake, *Land into Water,* 236.

50. Ken Woodburn, "A Brief History of Florida's Environmental Movement," http://www.myflorida.com/fdi/fscc/news/state/woodburn.htm.

51. Luther J. Carter, *The Florida Experience: Land and Water Policy in a Growth State* (Baltimore: Johns Hopkins University Press, 1974), 8.

52. Maurice O'Sullivan Jr. and Jack C. Lane, eds., *The Florida Reader: Visions of Paradise from 1530 to the Present* (Sarasota, Fla.: Pineapple Press, 1991), 224.

53. Audubon of Florida, time line, http://www.audubonofflorida.org/main/timeline.htm.

54. Nathaniel P. Reed, "Facing Florida's Growth: State Needs More Than 1000 Friends to Keep Eden from Vanishing Forever," speech to 1000 Friends of Florida, Lost Tree Village, Fla., November 18, 1998.

55. David E. Dodrill, *Selling the Dream: The Gulf American Corporation and the Building of Cape Coral, Florida* (Tuscaloosa: University of Alabama Press, 1993), 15–20.

56. Ibid., 31.

57. Don Ruane, "Cape Stiffens Water Rules," *Fort Myers News-Press,* March 1, 2005.

58. Ibid., 233.

59. Robert A. Catlin, *Land Use Planning, Environmental Protection and Growth Management: The Florida Experience* (Chelsea, Mich.: Ann Arbor Press, 1997), 43.

60. David R. Colburn and Lance deHaven-Smith, *Government in the Sunshine State: Florida Since Statehood* (Gainesville: University Press of Florida, 1999), 51–52.

61. Nathaniel P. Reed, interview by Julian Pleasants, University of Florida Samuel Proctor Oral History Program, November 2, 2000, and December 18, 2000, 7.

62. Jonathan Aitken, *Nixon: A Life* (Washington, D.C.: Regnery Publishing, 1993), 398.

63. Reed, interview, 16.

64. Carter, *The Florida Experience,* 54.

65. Ibid.

66. Edmund F. Kallina Jr., *Claude Kirk and the Politics of Confrontation* (Gainesville: University Press of Florida, 1993), 155.

67. Claude Roy Kirk Jr., interview by Julian Pleasants, University of Florida Samuel Proctor Oral History Program, October 29 1998, 138.

68. Richard E. Foglesong, *Married to the Mouse: Walt Disney World and Orlando* (New Haven: Yale University Press, 2001), 60.

69. Ibid., 62.

70. Ibid., 80.

71. Allen Morris, *The Florida Handbook, 2001–2002* (Tallahassee: Peninsular Publishing, 2001), 397.

72. Author interview with Tim O'Brien, editor, *Amusement Business News.*

73. Robert Johnson, "Dry Times at Epcot," *Orlando Sentinel,* September 1, 2001.

CHAPTER 2

1. From Frontinus, *De Aquaeductu,* translated in Harry B. Evans, *Water Distribution in Ancient Rome: The Evidence of Frontinus* (Ann Arbor: University of Michigan Press, 1994), 14.

2. A. Trevor Hodge, *Roman Aqueducts and Water Supply* (London: Gerald Duckworth and Co., 1992). See pp. 5–9 for discussion of aqueducts as necessity versus luxury; pp. 347–48 for aqueduct statistics.

3. Ibid., 5–9.

4. Ibid., 9.

5. "Estimated Use of Water in the United States, 2000," U.S. Geological Survey Circular 1268, updated February 2005, 1.

6. Ibid., 8.

7. *Water on Tap: What You Need to Know,* U.S. Environmental Protection Agency Office of Water, EPA 816-K-03–007, October 2003, 10, http://www.epa.gov/safewater.

8. Ibid.

9. "U.S. Per Capita Water Use Falls to 1950s Levels," press release, Pacific Institute, March 11, 2004, http://www.pacinst.org/press_center/usgs/.

10. "Estimated Use of Water in the United States, 2000."

11. Allen Morris and Joan Perry Morris, eds., *The Florida Handbook, 2005–2006* (Tallahassee: Peninsular Publishing, 2005), 360.

12. Morton D. Winsberg, *Florida Weather,* second edition (Gainesville: University Press of Florida, 2003), 1.

13. Ibid.

14. Elizabeth D. Purdum, *Florida Waters,* published jointly by Florida's water management districts, April 2002, 49.

15. "U.S. Residential Swimming Pool and Hot Tub Industry Shipments, 2004," Association of Pool and Spa Professionals.

16. Frontinus, translated in Evans, *Water Distribution in Ancient Rome,* 20.

17. Richard L. Marella and Marian P. Berndt, "Water Withdrawals and Trends from the Floridan Aquifer System in the Southeastern United States, 1950–2000," U.S. Geological Survey Circular 1278, 2005, 1.

18. Ibid., 16.

19. "Background Information on the Peace River Basin," Southwest Florida Water Management District paper, August 2004, 5.

20. Marella and Berndt, "Water Withdrawals and Trends," 17.

21. Ibid.

22. "Suwannee River Basin and Estuary Initiative: Executive Summary," U.S. Geological Survey open file report 2004–1198, April 2004, 3.

23. Centers for Disease Control, nitrate fact sheet, http://www.cdc.gov/ncidod/dpd/healthywater/factsheets/nitrate.htm.

24. Bruce Ritchie, "Scientists Worried about Wakulla: Drinking-Water Standards Don't Fit for Spring," *Tallahassee Democrat,* March 21, 2006.

25. Amanda Riddle, "Florida Drought over, but Water Shortage Still Critical," Associated Press, December 27, 2001.

26. "Getting Ready as Drought Worsens," *Tampa Tribune,* March 19, 2001.

27. Craig Pittman, "Parched: Wells and Waterways Are Running Dry around the State, but Experts Say the Worst of the Drought Is Yet to Come," *St. Petersburg Times,* March 9, 2001.

28. "Frequently Asked Questions about Water Restrictions," South Florida Water Management District, 2001, http://www.sfwmd.gov/curre/watshort/q_and_a2.html (last accessed April 2005).

29. "The Water Crisis Becomes Real," *Lakeland Ledger,* March 4, 2001.

30. Richard L. Marella, "Water Withdrawals, Use, Discharge and Trends in Florida, 2000," U.S. Geological Survey Scientific Investigations Report 2004–5151, 22.

31. Ibid., 1.

32. Ibid., 15. This figure is different from the "domestic use" figure cited earlier. Florida's average domestic per capita use was 106 gallons a day in 2000, slightly higher than the 103 gallons per day figure in 1995 and 16 gallons higher than the national domestic average of 90 gallons per day.

33. Ibid.

34. Ibid.

35. While the author interviewed Sol Koppel, this story was originally reported in the *St. Petersburg Times.* Justin George, "Home Owner's Xeriscaped Lawn Draws Wrath of Association," *St. Petersburg Times,* March 26, 2004.

36. National Golf Foundation 2004 count for eighteen-hole courses. Florida was number one with 1,218 courses, and California came in second with 906.

37. Marella, "Water Withdrawals, Use, Discharge and Trends," 28.

38. Florida Department of Environmental Protection, "Florida Water Conservation Initiative," April 2002, 9.

39. Peter H. Gleick, *The World's Water, 2004–2005: The Biennial Report on Freshwater Resources* (Washington, D.C.: Island Press, 2004), 10.

40. Paul Simon, *Tapped Out: The Coming World Crisis in Water and What We Can Do about It* (New York: Welcome Rain Publishers, 1998), 9.

41. Gleick, *The World's Water, 2004–2005,* 2.

42. Lester R. Brown, "How Water Scarcity Will Shape the New Century," speech, Stockholm Water Conference, August 14, 2000.

43. Gleick, *The World's Water, 2004–2005,* 1.

44. Mary H. Cooper, "Water Shortages: Is There Enough Fresh Water for Everyone?" *CQ Researcher* 13, no. 27 (2003): 652.

45. Author interview with Dr. Kiran Patel.

46. Fiona Harvey, "Liquid Asset? Why Drought Overshadows World Growth," *Financial Times,* March 21, 2006. This statistic is from the World Health Organization.

47. Ibid.

48. "Water for People, Water for Life—UN World Water Development Report," UNESCO, March 2003, 8.

49. Ibid., 9.

50. Purdum, *Florida Waters,* 37.

51. See "Florida's Water Cycle," in Purdum's *Florida Waters,* 38. The state receives 150 billion gallons of rainfall a day, and another 26 billion inflow from rivers to the north. Floridians consume 8.2 billion gallons of water a day. Add that to 107 billion gallons cycled out by evapotranspiration and the 66 billion gallons that flow into the sea, and losses now outweigh gains by 5.2 billion gallons a day.

52. Ibid.

CHAPTER 3

1. Julie Rhinehart, "The Grand Dam," U.S. Bureau of Reclamation, http://www.usbr.gov/lc/hooverdam/history/articles/rhinehart1.html.

2. The similarity to the western U.S. and North African climate can be found at "Distribution of Non-Polar Arid Lands," U.S. Geological Survey map, http://pubs.usgs.gov/gip/deserts/what/world.html.

3. 2005 National Inventory of Dams, http://crunch.tec.army.mil/nid/webpages/nid.cfm.

4. Marq de Villiers, *Water: The Fate of Our Most Precious Resource* (New York: Houghton Mifflin, 2000), 121.

5. Frederick Jackson Turner, "The Significance of the Frontier in American History," essay read at the meeting of the American Historical Association in Chicago, July 12, 1893. Accessed at the Modern History Sourcebook, Fordham University, http://www.fordham.edu/halsall/mod/1893turner.html.

6. Donald Worster, *Rivers of Empire: Water, Aridity and the Growth of the American West* (New York: Oxford University Press, 1985), 11.

7. Marc Reisner, *Cadillac Desert: The West and Its Disappearing Water* (New York: Penguin Books, 1993), 11.

8. Mike Obmascik, "Paradise Lost at Lake Foul," *Outside Online,* http://outside.away.com/outside/news/ozone/oz0722.html.

9. Tom Kenworthy, "Political Magic Blunts Budget Ax on Dam Project," *Washington Post,* June 29, 1995.

10. Ibid.

11. "Construction Begins on Long-Awaited Animas–La Plata Reservoir," *U.S. Water News Online,* August 2005, www.uswaternews.com/archives/arcsupply/5consbegi8.htm.

12. Reisner, *Cadillac Desert,* 308.

13. "Reclamation Facts and Information," http://www.usbr.gov/newsroom/presskit/factsheet/factsheetdetail.cfm?recordid=1.

14. Michael Parfit, "When Humans Harness Nature's Forces," in "Water: The Power, Promise and Turmoil of North America's Fresh Water," special edition, *National Geographic* 184, no. 5A (1993).

15. Richard Munson, "Why the Tennessee Valley Authority Must Be Reformed," Northeast Midwest Institute, September 2001; Reisner, *Cadillac Desert,* 324–28.

16. Richard Hamann, "Wetlands Loss in South Florida and the Implementation of Section 404 of the Clean Water Act," report to the Office of Technology Assessment Oceans and Environment Programs, Office of Technology Assessment, U.S. Congress, September 1982, 18–19.

17. Ibid., 19–20.

18. Robert P. King, "Katrina Galvanized Lake O Consultants," *Palm Beach Post,* May 21, 2006.

19. "Crossroads: Congress, the Corps of Engineers and the Future of America's Water Resources," National Wildlife Federation and Taxpayers for Common Sense, March 2004, 21.

20. Scott Faber, "Corps of Engineers May Get More Restoration Projects, Scrutiny," *Chesapeake Bay Journal,* September 2003.

21. "Crossroads," 67.

22. Jan Goldman-Carter, Andrew Schock, Nancy A. Payton, and Ann W. Hauck, "Road to Ruin: How the U.S. Government Is Permitting the Destruction of the Western Everglades," National Wildlife Federation, Florida Wildlife Federation, Council of Civic Associations, November 2002.

23. "Complaint for Declaratory and Injunctive Relief," *National Parks Conservation Association v. U.S. Army Corps of Engineers,* Case No. 1: 2006CV20256, U.S. District Court, Southern District of Florida, filed January 31, 2006.

24. Tom Henry, "The Future of the Great Lakes: Who Will Control the Water?" *Toledo Blade,* June 10, 2001.

25. Stephen Ohlemacher, "Growing Population Shifts Political Power," Associated Press, December 22, 2005.

26. Paul Simon, speech, U.S. Bureau of Reclamation Conference, "Growing the U.S. Water Supply through Purification Technologies," Golden, Colo., conference papers published May 2000, 7.

27. Reisner, *Cadillac Desert,* 308.

28. U.S. Bureau of the Census.

29. Bob Graham, interview by Samuel Proctor, University of Florida Samuel Proctor Oral History Program, May 3, 1991, 229–30.

30. Sydney P. Freedberg, "Loophole Inc.: A Special Report on Florida's Corporate Income Tax," *St. Petersburg Times,* October 26, 2003.

31. Reubin O'Donovan Askew, "Remarks of Reubin O'D. Askew Governor of Florida at the Governor's Conference on Water Management in South Florida," Miami Beach, Fla., September 22, 1971, 5. The underline was Askew's.

32. Catlin, *Land Use Planning, Environmental Protection and Growth Management,* 52.

33. Florida Statute 373.016.

34. Richard A. Pettigrew, interview by Julian Pleasants, University of Florida Samuel Proctor Oral History Program, May 23, 2001, 29.

35. U.S. Bureau of the Census.

36. "Threats to Wetlands," U.S. Environmental Protection Agency Office of Wetlands, EPA 843-F-01–002d, March 2002.

37. Hamann, "Wetlands Loss in South Florida," 26.

38. Cynthia Barnett, "Squeezed: Does Big Citrus Have a Future in Florida?" *Florida Trend,* March 2003, 46.

39. Marella, "Water Withdrawals, Use, Discharge and Trends," 17.

40. Debbie Salamone, "Paving It Over," part 5 in "Florida's Water Crisis," *Orlando Sentinel,* May 26, 2002.

41. Author interview with Bob Smith.

42. Theo D. Baars Jr., "Land Use without Zoning for Florida," in *The Use of Florida's Land: Summary Report of the Environmental Land Management Study Committee's Conference on Land Use* (Boca Raton: Florida Atlantic University—Florida International University Joint Center for Environmental and Urban Problems, June 1973), 30.

43. Charles J. Zwick, interview by Cynthia Barnett, University of Florida Samuel Proctor Oral History Program, August 17, 2000, 20.

44. Michael Grunwald, *The Swamp: The Everglades, Florida and the Politics of Paradise* (New York: Simon and Schuster, 2006), 271.

45. Mary Kelley Hoppe, "Charting the Course for the Bay's Recovery," *Bay Soundings,* Winter 2006.

46. Kate Spinner, "Hundreds Meet in Sanibel over Caloosahatchee Water Releases," *Bonita News,* November 11, 2005.

47. Hamann, "Wetlands Loss in South Florida," 27–38.

CHAPTER 4

1. Curtis H. Marshall, Roger A. Pielke Sr., Louis T. Steyaert, and Debra A. Willard, "The Impact of Anthropogenic Land-Cover Change on the Florida Peninsula Sea Breezes and Warm Season Sensible Weather," *Monthly Weather Review* 132 (January 2004): 28–52; and Curtis H. Marshall, Roger A. Pielke Sr., and Louis T. Steyaert, "Has the Conversion of Natural Wetlands to Agricultural Land Increased the Incidence and Severity of Damaging Freezes in South Florida?" *Monthly Weather Review* 132 (September 2004): 2243–58.

2. "The Impact of Anthropogenic Land-Cover Change"; and Rebecca Lindsey, "Deep Freeze and Sea Breeze: Changing Land and Weather in Florida," *NASA Earth Observatory,* May 17, 2005.

3. "The Impact of Anthropogenic Land-Cover Change," 47.

4. Ibid., 51.

5. Curtis H. Marshall, Roger A. Pielke Sr., and Louis T. Steyaert, "Crop Freezes and Land-Use Change in Florida," *Nature* 426 (November 2003).

6. Ibid.

7. Author interview with James O'Brien.

8. Author interview with Roger Pielke Sr.

9. Michael Wall, "Dr. Curry Goes to Washington," *Creative Loafing Atlanta,* November 2, 2005.

10. U.S. Environmental Protection Agency, "Global Warming—Climate," http://yosemite.epa.gov/oar/globalwarming.nsf/content/ClimateFutureClimateUSClimate.html.

11. Author interview with William H. Schlesinger.

12. Alan Cooperman, "Evangelicals Will Not Take Stand on Global Warming," *Washington Post,* February 2, 2006.

13. "Arctic Ocean Could Be Ice-Free in Summer within 100 Years, Scientists Say," University of Arizona press release, August 28, 2005.

14. Stefan Lovgren, "Warming to Cause Catastrophic Sea Level Rise?" *National Geographic,* April 26, 2004.

15. *Time,* October 3, 2005.

16. Ibid., 44.

17. Ted Steinberg, *Acts of God: The Unnatural History of Natural Disaster in America* (New York: Oxford University Press), 54.

18. Ibid., 67.

19. "Hurricane History," National Hurricane Center, http://www.nhc.noaa.gov/HAW2/english/history.shtml#donna.

20. Ibid., http://www.nhc.noaa.gov/HAW2/english/history.shtml#camille.

21. U.S. Census 2000, http://www.census.gov.

22. Analysis provided by Gary Appelson, Caribbean Conservation Corporation.

23. "Crossroads," 43.

24. Mike Williams, "Staying Ahead of the Storm," *Austin American-Statesman,* October 2, 2005.

25. "President Outlines Hurricane Katrina Relief Efforts," speech, White House Rose Garden, August 31, 2005.

26. R. H. Caffey and M. Schexnayder, "Coastal Louisiana and South Florida: A Comparative Wetland Inventory," Interpretive Topic Series on Coastal Wetland Restoration in Louisiana, National Sea Grant Library, 2003.

27. Howard T. Odum, Elisabeth C. Odum, and Mark T. Brown, *Environment and Society in Florida* (Boca Raton, Fla.: Lewis Publishers, 1997), 189.

28. "Importance of Wetlands," U.S. Environmental Protection Agency, http://www.epa.gov/bioindicators/aquatic/importance.html.

29. Odum, Odum, and Brown, *Environment and Society in Florida,* 185.

30. Sharon Begley, "Man-Made Mistakes Increase Devastation of 'Natural' Disasters," *Wall Street Journal,* September 2, 2005.

31. Glen Martin, "Wetland Restoration Seen as Crucial," *San Francisco Chronicle,* September 5, 2005.

32. Begley, "Man-Made Mistakes Increase Devastation."

33. S. V. Dáte, *Quiet Passion: A Biography of Senator Bob Graham* (New York: Tarcher/Penguin, 2004), 35–36.

34. Ibid., 158.

35. Ibid.

36. Douglas Hanks III, "Family Celebrates 70th Year of Empire Built on Miami Development," *Miami Herald,* January 25, 2002.

37. Florida Environmental Land Management Study Committee, "Final Report of the Environmental Land Management Study Committee," ELMS II, February 1984, 40.

38. Randall G. Holcombe, "Growth Management in Action: The Case of Florida," report to the Governor's Growth Management Study Commission, August 2000.

39. Robert Barnes, "Growth: It's the Year's Key Issue; Will Anything Be Done?" *St. Petersburg Times,* March 17, 1985.

40. Graham, interview, May 3, 1991; April 6, 1994; February 21, 1995; and January 26, 2001, 456.

41. Barnes, "Growth."

42. Zwick, interview, 21.

43. Grunwald, *The Swamp,* 256.

44. Ibid., 275–77.

45. Author interview with Lou Toth.

46. Ibid.

47. Florida Department of Environmental Protection, "Processing of Individual Wetland Resource Permits," http://www.dep.state.fl.us/water/wetlands/delineation/wet perm.htm (last accessed September 24, 2003).

48. Author interview with Donna Blanton.

49. Mike Vogel, "Sprawling Contradiction," *Florida Trend,* December 2001.

50. John deGroot and Chuck Clark, "The 'Burb That Ate the Wetlands," *Fort Lauderdale Sun-Sentinel,* September 15, 1991.

51. Ibid.

52. Ibid.

53. Vogel, "Sprawling Contradiction."

CHAPTER 5

1. This story was recounted by Reed in the oral history interview first cited in chap. 1. See Reed, interview, pp. 49 and 50.

2. EPA time line, "30 Years of Environmental Progress," http://www.epa.gov/region3/annual_report/PDFs/TimeLine.pdf.

3. J. Brooks Flippen, *Nixon and the Environment* (Albuquerque: University of New Mexico Press, 2000), 47.

4. Aitken, *Nixon,* 397–98.

5. Flippen, *Nixon and the Environment,* 47.

6. The Fortune 500, 2005.

7. Donald Worster, *Nature's Economy: A History of Ecological Ideas* (Cambridge: Cambridge University Press, 1994), 273.

8. Colburn and deHaven-Smith, *Government in the Sunshine State,* 13, 31.

9. Author interview with David Colburn.

10. Charles W. Tebeau, *A History of Florida* (Miami: University of Miami Press, 1971), 449.

11. Wade L. Hopping, interview by Cynthia Barnett, University of Florida Samuel Proctor Oral History Program, July 22, 2003, 25.

12. Alexander P. Lamis, *The Two Party South* (New York: Oxford University Press, 1984), 179–92.

13. Author interview with Colburn.

14. Colburn and deHaven-Smith, *Florida's Megatrends,* 56; and Vogel, "Good Migrations."

15. Alecia Swasy and Robert Trigaux, "Make the Money and Run," *St. Petersburg Times,* September 20, 1998.

16. Andres Viglucci and Alfonso Chardy, "Bush and Business: Fast Success, Brushes with Mystery," *Miami Herald,* October 5, 2002.

17. "A Test for Bush," *St. Petersburg Times,* January 10, 1999.

18. Julie Hauserman and Craig Pittman, "An Unnatural Silence," *St. Petersburg Times,* October 14, 2002.

19. Ron Matus, "A History of Trouble," *Gainesville Sun,* June 13, 1999.

20. Hauserman and Pittman, "An Unnatural Silence."

21. Jim McGee, Peter Wallsten, and James Grotto, "Paving Firm Gave $190,000 to GOP Accounts after Florida Deal Was Sealed," *Miami Herald,* October 27, 2002.

22. Author interview with Jeb Bush.

23. Christine Todd Whitman, *It's My Party Too: The Battle for the Heart of the GOP and the Future of America* (New York: Penguin Press, 2005), 150.

24. Thomas E. Dahl, "Wetlands Losses in the United States, 1780's to 1980's," U.S. Department of the Interior, Fish and Wildlife Service, 1990.

25. Cynthia Barnett, "Swamp Repair: Florida's Latest Weapon in the Battle to Preserve Its Dwindling Wetlands Isn't Living Up to Its Promise," *Florida Trend,* May 2000.

26. Gale Norton press conference, Department of the Interior, Washington D.C., March 30, 2006.

27. As just one example, see Florida State University professor J. B. Ruhl's "The Effects of Wetlands Mitigation Banking on People," FSU College of Law Public Research Paper No. 179, January 2006.

28. Thomas E. Dahl, "Status and Trends of Wetlands in the Conterminous United States, 1998 to 2004," U.S. Department of the Interior, Fish and Wildlife Service, 2006. See pp. 74 to 77 for discussion and maps of human-made ponds.

29. "Ponds Proliferate, but Wetlands Losses Continue," press release, Association of State Wetland Managers, March 30, 2006.

30. Craig Pittman and Matthew Waite, "Vanishing Wetlands: They Won't Say No," *St. Petersburg Times,* May 22, 2005.

31. Ibid.

32. Goldman-Carter, Schock, Payton, and Hauck, "Road to Ruin."

33. Pittman and Waite, "Vanishing Wetlands."

34. Florida Home Builders Association.

35. Kathryn Ziewitz and June Wiaz, *Green Empire: The St. Joe Company and the Remaking of Florida's Panhandle* (Gainesville: University Press of Florida, 2004) 5, 42–43.

36. Ibid., 131–33.

37. Craig Pittman, "Brought to You by the St. Joe Company (with Your Help)," *St. Petersburg Times,* April 21, 2002.

38. Craig Pittman, "State Park Stands in Road's Way," *St. Petersburg Times,* July 13, 2003.

39. Robert Trigaux, "Panhandle, Make Way for St. Joe Co.'s Great Northwest," *St. Petersburg Times,* March 25, 2001.

40. "DEP Signs Panhandle Wetlands Permitting Deal with St. Joe," Associated Press, February 25, 2004.

41. Craig Pittman, "Development vs. Environment Leads to Give-and-Take Meeting," *St. Petersburg Times,* April 21, 2002.

42. Craig Pittman and Matthew Waite, "Judge Halts St. Joe Development," *St. Petersburg Times,* November 11, 2005.

43. Peter Rummell, "The St. Joe Co. 2004 Annual Report," 6.

44. The author's original reporting on conservation easements was for the cover story "Uncertain Legacy," *Florida Trend,* August 2004. Updated conservation easement data from the Florida Department of Environmental Protection.

45. Barnett, "Uncertain Legacy."

46. Ibid.

47. Craig Pittman and Joni James, "Biggest Florida Land Buy Planned," *St. Petersburg Times,* November 22, 2005.

48. Ibid.

49. Author interview with Bob Smith.

50. Grunwald, *The Swamp,* 2.

51. The author's original reporting on the Comprehensive Everglades Restoration Plan was for the cover story "Fragile," cowritten with Mike Vogel, *Florida Trend,* March 1, 2001.

52. Michael Grunwald, "A Rescue Plan, Bold and Uncertain," *Washington Post,* June 23, 2002.

53. Michael Grunwald, "Florida Steps in to Speed Up State-Federal Everglades Cleanup," *Washington Post,* October 14, 2004.

54. Flippen, *Nixon and the Environment,* 228.

55. Hopping, interview.

56. Michael Shellenberger and Ted Nordhaus, "The Death of Environmentalism," October 2004, available at www.thebreakthrough.org/images/Death_of_Environmental ism.pdf.

57. Author interview with Eric Draper.

58. "Testimony of Dr. Peter H. Gleick, President, Pacific Institute," Senate Commit tee on Natural Resources and Water, February 14, 2006.

59. Author interview with Paula Dockery.

CHAPTER 6

1. Robert Trigaux, "Developer's Recent Hot Streak Carries Over into IPO Market," *St. Petersburg Times,* March 15, 2002, 1-E. Hoffman told Trigaux he was "thrilled" by the success of WCI's IPO.

2. WCI Communities promotional materials and press releases, http://www.wci communities.com/.

3. Mike Seemuth, "Heading for Higher Ground," *Florida Trend,* December, 1996.

4. "Embassy Row," Center for Responsive Politics, www.opensecrets.org/bush/ambassadors/hoffman.asp.

5. "Florida's Influential," *Florida Trend,* November 2004.

6. Michael Grunwald, "Growing Pains in Southwest Florida: More Development Pushes Everglades to the Edge," *Washington Post,* June 25, 2002.

7. Peter Wallsten, "Bush Brothers, State Republicans Celebrate, Increase Party's Wealth," *Miami Herald,* June 22, 2002.

8. Thomas, Warren, and Associates, "The Impacts of Florida's Mature Residents," Phoenix, Ariz., May 1, 2002.

9. "RE: Economic Impact of the 50+ Market," memorandum from Al Hoffman to Governor Jeb Bush and Kathleen Shanahan, June 10, 2002, 2 (emphasis in original).

10. "Florida, California and Texas to Dominate Future Population Growth." An attached chart shows Florida's population as 17,509,827 in 2005 and 21,204,132 in 2015, a 21.1 percent increase.

11. Colburn and deHaven Smith, *Florida's Megatrends,* 84.

12. "Governor Bush Announces Creation of *Destination Florida* Commission," press release, Office of the Governor, July 29, 2002.

13. The author originally reported the Destination Florida story in "How Many Retirees?" *Florida Trend,* February 2003.

14. "Governor Bush Announces Creation of *Destination Florida* Commission."

15. "How Many Retirees?"

16. E-mail from Barbara Matthews of Tampa to Governor Jeb Bush, July 30, 2002.

17. "How Many Retirees?"

18. Kirk Johnson, "Drought Rules End, but They May Come Again Another Day," *New York Times,* October 31, 2002.

19. Yvette C. Hammett, "Water Use Closes in on Permit Limits," *Tampa Tribune,* September 28, 2002.

20. "Council History," Florida Council of 100, http://www.fc100.org.

21. Craig Pittman and Julie Hauserman, "North Has It, South Wants It," *St. Petersburg Times,* August 10, 2003.

22. "Improving Florida's Water Supply Management Structure: Ensuring and Sustaining Environmentally Sound Water Supplies and Resources to Meet Current and Future Needs," Florida Council of 100, September 2003, 3.

23. Ibid., 2.

24. E-mail from Lee Arnold to Governor Jeb Bush, January 29, 2003.

25. "Improving Florida's Water Supply Management Structure," 23.

26. Pittman and Hauserman, "North Has It, South Wants It."

27. U.S. Bureau of the Census, "Persons per Square Mile 2000," www.census.gov.

28. Buddy Blain, interview by Julian Pleasants, University of Florida Samuel Proctor Oral History Program, February 17, 2003.

29. Honey Rand, *Water Wars: A Story of People, Politics and Power* (Philadelphia: Xlibris, 2003), 24.

30. Ibid., 61–62.

31. Ibid., 68.

32. Ibid., 183.

33. Ibid., 229.

34. Author interview with Lee Arnold.

35. Craig Pittman and Lucy Morgan, "Fierce Outcry Sidelines Water Distribution Plan," *St. Petersburg Times,* January 22, 2004.

36. Richard Hamann, "Law and Policy in Managing Water Resources," in *Water Resources Atlas of Florida,* ed. Edward A. Fernald and Elizabeth D. Purdum (Tallahassee: Florida State University Institute of Science and Public Affairs, 1998), 307.

37. Curtis Morgan, "North Florida Water-Use Plan Faces Opposition," *Miami Herald,* October 15, 2003.

38. National Hurricane Center building brochure.

39. Julie Hauserman, "The Candidates' Stands on Governing Growth," *St. Petersburg Times,* October 25, 1998.

40. Pittman and Morgan, "Fierce Outcry Sidelines Water Distribution Plan."

41. Destination Florida Commission, "Final Report," February 2003, 46.

42. Ibid., 45.

43. U.S. District Judge Donald Middlebrooks, Southern District of Florida, Order on Cross Motions for Summary Judgment, September 30, 2005, 36.

44. Todd Hartman, "Dividing the Waters," *Rocky Mountain News,* July 11, 2003.

CHAPTER 7

1. Upper Chattahoochee Riverkeeper, http://www.ucriverkeeper.org/Ourriver5.htm.

2. Georgia Department of Natural Resources, Environmental Protection Division, "Chattahoochee River Basin Watershed Protection Plan," 1997, 3–11.

3. Consent decree, *Upper Chattahoochee Riverkeeper, Inc., et al. v. City of Atlanta,* Case No. 1: 95-CV-2550-FMH (N.D. GA 1997).

4. Ibid.

5. Joe Cook and Monica Cook, *River Song: A Journey down the Chattahoochee and Apalachicola Rivers* (Tuscaloosa: University of Alabama Press, 2000), 258.

6. Ibid.

7. "Metro Atlanta Water Resources Summary," Metro Atlanta Chamber of Commerce.

8. Southern Environmental Law Center, http://www.southernenvironment.org/cases/atlanta/facts.htm.

9. "Smog FAQ," *Atlanta Journal-Constitution* online, http://www.ajc.com/news/content/health/smog/faq.html.

10. Michael Ludden, "Downtown's Making a Comeback," *Atlanta,* October 2005.

11. Bruce Ritchie, "Atlanta Officials Fight to Use More River Water for City's Needs," *Tallahassee Democrat,* November 5, 2001.

12. Figure from Robert Kerr, Georgia's negotiator in the Apalachicola-Chattahoochee Flint River Basin Compact.

13. Douglas Jehl, "Atlanta's Growing Thirst Creates Water War," *New York Times,* May 27, 2002.

14. De Villiers, *Water,* 16.

15. Personal communication with Barbara Schmidt.

16. Donald Worster, *A River Running West: The Life of John Wesley Powell* (New York: Oxford University Press, 2001), 357.

17. Robert Glennon, "Water Scarcity, Marketing and Privatization," in "Symposium of Waterbanks, Piggybanks and Bankruptcy: Changing Directions in Water Law," *Texas Law Review,* June 2005, 1898.

18. "Estimated Use of Water in the United States, 2000," 1; http://www.census.gov; and William M. Alley, Thomas E. Reilly, and O. Lehn Franke. "Sustainability of Ground-Water Resources," U.S. Geological Survey Circular 1186, 1999, 1–2.

19. Douglas Jehl, "A New Frontier in Water Wars Emerges in East," *New York Times,* March 3, 2003.

20. Laura Cadiz, "A Battle over Treasured Waters," *Baltimore Sun,* November 24, 2002.

21. J. B. Ruhl, "Equitable Apportionment of Ecosystem Services: New Water Law for a New Water Age," *Journal of Land Use and Environmental Law* 47 (Fall 2003): 47.

22. Henderson, "Who Gets the Water?"

23. David H. Getches, *Water Law in a Nutshell* (St. Paul, Minn.: West Publishing, 1997), 397; and Ruhl, "Equitable Apportionment of Ecosystem Services," 49.

24. Ruhl, "Equitable Apportionment of Ecosystem Services," 49.

25. Author interview with J. B. Ruhl.

26. Cooper, "Water Shortages," 660.

27. Steven Leitman, "Apalachicola-Chattahoochee-Flint Basin," in *Adaptive Governance and Water Conflict: New Institutions for Collaborative Planning,* ed. John T. Scholz and Bruce Stiftel (Washington, D.C.: Resources for the Future, 2005), 87.

28. Robert Kerr comments, "Association of American Law Schools Conference, Section on Natural Resources in Atlanta, Georgia," *Georgia State University Law Review,* Winter 2004, 257.

29. Bruce Ritchie, "New Volleys in Water Battle," *Tallahassee Democrat,* July 25, 2006.

30. Stacy Shelton, "No Extra Water for Florida from State," *Atlanta Journal-Constitution,* July 26, 2006.

31. Charles Lane, "Battle for Potomac Back in Court," *Washington Post,* October 7, 2003.

32. Robert P. King, "Battle Lines Form over Water Law Rewrite," *Palm Beach Post,* September 22, 2003.

33. Ibid.

34. Linda Greenhouse, "Supreme Court Backs Virginia in Rift over Potomac Water," *New York Times,* December 10, 2003.

35. Barlow Burke, "Association of American Law Schools Conference, Section on Natural Resources in Atlanta, Georgia," *Georgia State University Law Review,* Winter 2004, 290.

36. Joseph W. Dellapenna, "Interstate Struggles Over Rivers: The Southeastern States and the Struggle over the 'Hooch,'" in "Transboundary Water in the Twenty-First Century," *New York University Environmental Law Journal* 12, no. 3 (2003–2005): 851.

37. Tom Arrandale, "Beyond the Basin: Many States Are Revisiting Their Laws That Govern How Water Is Allocated," *Governing Magazine,* December 2004.

38. "The Great Lakes: An Environmental Atlas and Resource Book," U.S. Environmental Protection Agency, http://www.epa.gov/glnpo/atlas/index.html.

39. Arrandale, "Beyond the Basin."

40. Great Lakes Information Network, http://www.great-lakes.net/lakes/superior.html#overview.

41. James M. Olson, "Great Lakes Water: Limitations on Privatization and Diversions," Michigan Citizens for Water Conservation.

42. Water Resources Development Act of 1986, 1962d-20, Prohibition on Great Lakes Diversions.

43. Alan M. Schwartz, "The Canada-U.S. Environmental Relationship at the Turn of the Century," *Canadian Business and Current Affairs,* June 2000.

44. "It's Clear Ontario Water for Sale," *Toronto Star,* editorial, December 19, 2000.

45. Schwartz, "The Canada-U.S. Environmental Relationship at the Turn of the Century."

46. Glennon, "Water Scarcity, Marketing and Privatization," 1894.

47. Schwartz, "The Canada-U.S. Environmental Relationship at the Turn of the Century."

48. Peter H. Gleick, "Bagged and Dragged," *Scientific American* 284, no. 2 (2001): 53.

49. Ibid.

50. Author interview with Terry Spragg.

51. Sandra L. Postel and Aaron T. Wolf, "Dehydrating Conflict," *Foreign Policy* 126 (September–October 2001): 60.

52. Gleick, *The World's Water, 2004–2005,* 234.

53. Simon, *Tapped Out,* 59.

54. Felicity Barringer, "Growth Stirs a Battle to Draw More Water from the Great Lakes," *New York Times,* August 12, 2005.

55. Dan Egan, "Water Pressures Divide a Great Lake State," *Milwaukee Journal-Sentinel,* November 23, 2003.

56. Lee Bergquist, "Leaders Sign Accord on Great Lakes Water," *Milwaukee Journal-Sentinel,* December 13, 2005.

57. Lee Bergquist, "The Big Exception," *Milwaukee Journal-Sentinel,* November 29, 2003.

CHAPTER 8

1. Author interview with Rob Fisher, director, supply chain operations, Nestlé Waters North America water-bottling facility, Madison County, Fla. Most of the data and interviews in this chapter came from reporting associated with the author's "Looking for Clarity: A Report on the Bottled Water Industry in Florida," *Florida Trend,* February 2006.

2. Suwannee River Water Management District permit database.

3. Beverage Marketing Corporation.

4. Ibid.

5. Gleick, *The World's Water, 2004–2005,* 27–32.

6. Author interview with Kent Kise, director of quality and technical services, DS Waters of America.

7. Tom Standage, "Bad to the Last Drop," *International Herald Tribune,* August 2, 2005.

8. U.S. Bureau of the Census.

9. Jo Sandin, "Rancor That Flows like Water," *Milwaukee Journal-Sentinel,* October 14, 2000.

10. Author interview with Fisher.

11. Andrew Guy and Patty Cantrell, *Liquid Gold Rush* (Michigan Land Use Institute, 2001).

12. Hon. Lawrence C. Root, "Opinion Following Bench Trial," Case No. 01–14563-CE, 49.

13. David Eggert, "Bottled-Water Company Sues State over Restrictions," Associated Press State and Local Wire, June 17, 2005.

14. Granholm for Governor, http://www.granholmforgov.com/.

15. "Michigan's Job-Loss Streak Will Hit Six Years in 2006," press release, University of Michigan News Service, November 18, 2005.

16. Root, "Opinion Following Bench Trial," 66.

17. Robert Trigaux, "Florida Pulls Away from Nation on Job Growth," *St. Petersburg Times,* April 11, 2005.

18. Caroline Alexander, *The Way to Xanadu: Journeys to a Legendary Realm* (New York: Alfred A. Knopf, 1994).

19. Quoted in ibid., 65.

20. Ibid., 66.

21. Michael Dovaz, *Water: From the Big Bang to the Bottle* (New York: Assouline Publishing, 2000), 22.

22. James Call and Frank Stephenson, "Springtime in Florida," *Florida State University Research in Review,* Spring 2003.

23. G. G. Phelps and Stephen J. Walsh, "Characterization of the Hydrology, Water Quality and Aquatic Communities of Selected Springs in the St. Johns River Water Management District, Florida," U.S. Geological Survey and Florida Integrated Science Center, May 2005.

24. "Is Bottled Water Better Than Tap?" *20/20,* ABC News, May 6, 2005.

25. Gleick, *The World's Water, 2004–2005,* 23.

26. Ibid., 24.

27. Tara Boldt-Van Rooy, "'Bottling Up' Our Natural Resources: The Fight over Bottled Water Extraction in the United States," *Journal of Land Use* 18, no. 2 (2003): 279–81.

28. Guy and Cantrell, *Liquid Gold Rush.*

29. Letter from Brad Willis to Governor Jeb Bush, November 1, 2000.

CHAPTER 9

1. The City of West Palm Beach provides water to the town of Palm Beach. Analysis by City of West Palm Beach Management Information Systems in response to public records request, November 2004.

2. City of Belle Glade Water Department. Rate as of November 2004.

3. Analysis by Palm Beach County Water Utilities Department. See comparison graph at http://www.pbcwater.com/faq.htm.

4. Ibid.

5. "Avoiding Rate Shock: Making the Case for Water Rates," American Water Works Association, 2004, 9.

6. "The Poor Pay Much More for Water, Use Much Less—Often Contaminated," press release, World Water Council, 1999.

7. See chap. 2.

8. Water Partners International, http://www.water.org.

9. Congressional Budget Office, "Future Investment in Drinking Water and Wastewater Infrastructure," November 2002.

10. Steve Maxwell, "Water Is Cheap—Ridiculously Cheap!" *Journal of the American Water Works Association,* June 2005.

11. Ibid.

12. Adam Smith, *An Inquiry into the Nature and Causes of the Wealth of Nations* (New York: Modern Library, 1937), bk. 1, 28.

13. U.S. Department of Education, National Center for Education Statistics, http://nces.ed.gov/programs/quarterly/vol_6/6_4/8_3.asp/.

14. Author interview with Sheila Cavanagh Olmstead.

15. "Report Card for America's Infrastructure, 2005," American Society of Civil Engineers, http://www.asce.org/reportcard/2005/page.cfm?id=109#drinkingwater.

16. G. Tracy Mehan III, "Everyone Undervalues the True Worth of Water," *Detroit News,* June 19, 2003.

17. Ibid.

18. Darwin C. Hall, "Public Choice and Water Rate Design," in *The Political Economy of Water Pricing Reforms,* ed. Ariel Dinar (New York: Oxford University Press, 2000), 189.

19. President Bush State of the Union Address, January 29, 2002, http://www.whitehouse.gov/news/releases/2002/01/20020129–11.html/.

20. Julius Ciaccia, director, Cleveland Division of Water, "Statement on Financing Water Infrastructure Projects," testimony before the Water Resources and Environment Subcommittee on Transportation and Infrastructure, U.S. House of Representatives, June 8, 2005.

21. Author interview with Sanford V. Berg.

22. Author interview with Sanford V. Berg.

23. Garrett Hardin, "The Tragedy of the Commons," *Science* 162, no. 3859 (1968): 1244.

24. "The Value of Water," 189–90.

25. Ibid., 189.

26. Ibid., 197–201.

27. Ibid., 213.

28. "Lawmakers Learn about Sustainability," *Arizona Water Resource,* January–February 2005.

29. Author interview with Katie McCain, past president, American Water Works Association.

30. WaterTech Online, August 2003.

31. Author interview with Katie McCain.

32. John B. Whitcomb, "Florida Water Rates Evaluation of Single-Family Homes," July 13, 2005.

33. Sandra Postel, *Last Oasis: Facing Water Scarcity* (New York: W. W. Norton and Co., 1992), 166.

34. Peter H. Gleick, "A Soft Path: Conservation, Efficiency and Easing Conflicts over Water," in *Whose Water Is It? The Unquenchable Thirst of a Water-Hungry World,* ed. Bernadette McDonald and Douglas Jehl (Washington, D.C.: National Geographic Society, 2003), 188.

CHAPTER 10

1. Alley, Reilly, and Franke, "Sustainability of Ground-Water Resources," 26–28.

2. Katherine Yung, "Boone Pickens Dealing Again, but This Time the Commodity Is Water," *Dallas Morning News,* November 3, 2002.

3. Ibid.

4. "Mesa Water Releases Details of Water Purchase Initiative," Mesa Water news release, May 11, 2005.

5. Figures provided to the author by Mesa Water attorney Molly Cagle.

6. David Whitford, "Liquid Assets: Boone Pickens, Once an Oil Baron, Now Finds Water Cooler," *Fortune Small Business,* January 2006.

7. Interviews and reporting related to Azurix efforts to privatize water in Florida originally appeared in an article by the author, "Making Waves," *Florida Trend,* September 2000.

8. Michael Pollick and Chris Davis, "Enron's Grab for Florida's Water Was Factor in Collapse," *Sarasota Herald-Tribune,* August 15, 2002.

9. Ibid.

10. Maude Barlow, "The World's Water: A Human Right or a Corporate Good?" in McDonald and Jehl, *Whose Water Is It?* 38–39.

11. Postel, *Last Oasis,* 165.

12. "Water Allocation Markets," Florida Public Service Commission, September 2001.

13. Author interview with Michael Molligan.

14. Author interview with Mark D. Farrell.

15. Fredrik Segerfeldt, *Water for Sale: How Business and the Market Can Resolve the World's Water Crisis* (Washington, D.C.: Cato Institute, 2005), 33.

16. Barlow, "The World's Water," 33–34.

17. *The Water Barons: How a Few Powerful Companies Are Privatizing Your Water,* International Consortium of Investigative Journalists (Washington, D.C.: Public Integrity Books, 2003), 2.

18. Ibid.

19. "The Flood Dries Up," *Economist,* August 28, 2004.

20. Ralph Atkins, "RWE to Sell Off UK, US Water Interests," *Financial Times,* November 5/6 2005.

21. Douglas Jehl, "As Cities Move to Privatize Water, Atlanta Steps Back," *New York Times,* February 20, 2003.

22. Ibid.

23. Jon Luoma, "Water for Profit," *Mother Jones,* November/December 2002.

24. Scott Wallsten and Katrina Kosec, "Public or Private Drinking Water? The Effects of Ownership and Benchmark Competition on U.S. Water System Regulatory Compliance and Household Water Expenditures," AEI-Brookings Joint Center for Regulatory Studies, March 2005.

25. Ibid.

26. Segerfeldt, *Water for Sale,* 39.

27. "Water Privatization: A Broken Promise," Public Citizen's Critical Mass Energy and Environment Program, October 2001.

28. Jehl, "As Cities Move to Privatize Water, Atlanta Steps Back."

29. Craig Anthony (Tony) Arnold, "Privatization of Public Water Services: The States' Role in Ensuring Public Accountability," *Pepperdine Law Review* 32 (2005): 564.

30. Sanford V. Berg, "Survey of Benchmarking Methodologies," Executive Summary, February 2006.

31. Gary H. Wolff and Meena Palaniappan, "Public or Private Water Management? Cutting the Gordian Knot," *Journal of Water Resources Planning and Management,* January/February 2004.

CHAPTER 11

1. Cathy Keen, "Florida's Population Growth Little Affected by Last Year's Hurricanes," University of Florida News Desk, October 31, 2005.

2. Florida Department of Environmental Protection estimate.

3. See Senate Bill 1016, Desalination Water Supply Shortage Prevention Act of 2005, introduced by U.S. Senator Mel Martinez, Republican of Florida.

4. "Water Desalination Facilities, Energy and Reclaimed Water," Chairman Pete V. Domenici testimony, Senate Energy and Natural Resources Committee, October 20, 2005.

5. "Desalination and Water Purification Technology Roadmap," U.S. Bureau of Reclamation and Sandia National Laboratories, January 2003, 5.

6. Gleick, "A Soft Path," 192.

7. "Desalination and Water Purification Technology Roadmap," 3.

8. Peter Rogers, "Water Resources in the Twentieth and One Half Century: 1950–2050," Water Resources Update: Reflections on a Century of Water Science and Policy 116, March 2000.

9. Patrick Danner, "Successor to Sell Off the Last of Miami's General Development Corp.," *Miami Herald,* February 6, 2003.

10. Ben Weider and John Harry Fournier, "Activation Analyses of Authenticated Hairs of Napoleon Bonaparte Confirm Arsenic Poisoning," *American Journal of Forensic Medicine and Pathology* 20, no. 4 (1999): 378–82.

11. "Arsenic and Ground Water," American Ground Water Trust, 2005.

12. For an interesting political discussion of the new standard for arsenic, see Whitman, *It's My Party Too,* 157–61.

13. Roy E. Price and Thomas Pichler, "Arsenic and ASR in Southwest Florida: Source, Abundance, and Mobilization Mechanism Suwannee Limestone, Upper Floridan Aquifer," poster, Department of Geology, University of South Florida.

14. ASR V Conference, sponsored by the American Ground Water Trust, Tampa, Fla., October 20, 2005.

15. Author interview with Patrick Lehman.

16. "DNR Won't Budge on Aquifer Tests," *Green Bay Press-Gazette,* September 19, 2003.

17. National Research Council, *Regional Issues in Aquifer Storage and Recovery for Everglades Restoration: A Review of the ASR Regional Study Project Management Plan of the Comprehensive Everglades Restoration Plan* (Washington, D.C.: National Academies Press, 2002), 1.

18. Robert P. King, "Saltwater Conversion Possible Everglades Fallback," *Palm Beach Post,* February 1, 2005.

19. The author thanks Harry Smid, park services specialist, Fort Zachary Taylor Park, for tracking down the patent for the Marine Aerated Fresh Water Apparatus.

20. Peter H. Gleick, *The World's Water, 2000–2001: The Biennial Report on Freshwater Resources* (Washington, D.C.: Island Press, 2000), 93.

21. Postel, *Last Oasis,* 45.

22. Ibid., 46.

23. Douglas Jehl, "Alchemy or Salvation? Desalting the Sea," in McDonald and Jehl, *Whose Water Is It?* 206.

24. Gleick, *The World's Water, 2000–2001,* 96.

25. Jehl, "Alchemy or Salvation?" 207–8.

26. James E. Mielke, "Desalination R&D: The New Federal Program," in *Water Resources Issues and Perspectives,* ed. Ervin L. Clarke (New York: Nova Science Publishers, 2002), 161.

27. Estimate from Ken Herd, director of operations, Tampa Bay Water.

28. Craig Pittman, "Finish of Desal Plant Is Delayed," *St. Petersburg Times,* January 21, 2003.

29. Craig Pittman, "Cost to Fix Desalination Plant Jumps by Millions," *St. Petersburg Times,* August 7, 2004.

30. Author interview with Ken Herd.

31. King, "Saltwater Conversion Possible Everglades Fallback."

CHAPTER 12

1. Tom Shroder and John Barry, *Seeing the Light: Wilderness and Salvation; A Photographer's Tale* (New York: Random House, 1995), 37–41.

2. Author interview with Clyde and Niki Butcher. See Shroder and Barry, *Seeing the Light,* for the Butchers' tent-trailer camping and other adventures.

3. Shroder and Barry, *Seeing the Light,* 54.

4. "Who is the Next Ansel Adams?" *Popular Photography,* July 2004.

5. Nash, *Wilderness and the American Mind,* 326–27.

6. Author interview with Robert Carpenter.

7. Audubon of Florida, "Everglades Report," Summer 2005.

8. "Estimated Use of Water in the United States, 2000."

9. Craig Pittman and Matthew Waite, "Army Engineers Block Golf Course," *St. Petersburg Times,* December 8, 2005.

10. Ibid.

11. Nathaniel P. Reed comments at 2006 meeting of the Everglades Coalition.

12. Segerfeldt, *Water For Sale,* 16.

13. "The Soft Path for Water," Rocky Mountain Institute, http://www.rmi.org/sitepages/pid278.php.

14. Gleick, "A Soft Path," 189.

15. Sandra Postel, "Liquid Assets: The Critical Need to Safeguard Freshwater Ecosystems," Worldwatch Paper 170, July 2005.

16. Gleick, "A Soft Path," 189.

17. Tamara Lush, "Sure, the Storm Was Bad, But . . . ," *St. Petersburg Times,* November 5, 2005.

18. Ibid.

19. Tamara Lush, Abhi Raghunathan, Steve Bousquet, and Steve Thompson, "Was South Florida Ready for Wilma?" *St. Petersburg Times,* October 27, 2005.

20. Ibid.

21. Nash, *Wilderness and the American Mind,* 151.

22. Ibid., 150.

23. Ibid.

24. Richard Louv, *Last Child in the Woods: Saving Our Children from Nature-Deficit Disorder* (Chapel Hill: Algonquin Books, 2006), 34.

25. Lush, "Sure, the Storm Was Bad, But . . ."

26. For information on the Water Drop Patch developed for the Girl Scouts by the EPA, see http://www.epa.gov/adopt/patch/html/require.html.

Bibliography

Aitken, Jonathan. *Nixon: A Life.* Washington, D.C.: Regnery Publishing, 1993.

Alexander, Caroline. *The Way to Xanadu: Journeys to a Legendary Realm.* New York: Alfred A. Knopf, 1994.

Alley, William M., Thomas E. Reilly, and O. Lehn Franke. "Sustainability of Ground-Water Resources." U.S. Geological Survey Circular 1186, 1999.

Askew, Reubin O'Donovan. "Inaugural Address of Reubin Askew as Governor of Florida." Tallahassee, Fla., January 5, 1971.

———. Interview by Julian Pleasants, University of Florida Samuel Proctor Oral History Program, May 8, 1998.

———. "Remarks of Reubin O'D. Askew Governor of Florida at the Governor's Conference on Water Management in South Florida." Miami Beach, Fla., September 22, 1971.

"Avoiding Rate Shock: Making the Case for Water Rates." American Water Works Association, April 2004.

Baumann, Duane D., John J. Boland, and W. Michael Hanemann. *Urban Water Demand Management and Planning.* New York: McGraw-Hill, 1997.

Blain, Buddy. Interview by Julian Pleasants, University of Florida Samuel Proctor Oral History Program, February 17, 2003.

Blake, Nelson Manfred. *Land into Water—Water into Land: A History of Water Management in Florida.* Tallahassee: University Press of Florida, 1980.

Burt, Al. *Al Burt's Florida: Snowbirds, Sand Castles and Self-Rising Crackers.* Gainesville: University Press of Florida, 1997.

Carter, Luther J. *The Florida Experience: Land and Water Policy in a Growth State.* Baltimore: Johns Hopkins University Press, 1974.

Clarke, Ervin L., ed. *Water Resources Issues and Perspectives.* New York: Nova Science Publishers, 2002.

Colburn, David R., and Lance deHaven-Smith. *Florida's Megatrends: Critical Issues in Florida.* Gainesville: University Press of Florida, 2002.

———. *Government in the Sunshine State: Florida Since Statehood.* Gainesville: University Press of Florida, 1999.

Comprehensive Everglades Restoration Plan (CERP). http://www.evergladesplan.org.

Cook, Joe, and Monica Cook. *River Song: A Journey down the Chattahoochee and Apalachicola Rivers.* Tuscaloosa: University of Alabama Press, 2000.

Cooper, Mary H. "Water Shortages: Is There Enough Fresh Water for Everyone?" *CQ Researcher* 13, no. 27 (2003): 649–72.

Cowdrey, Albert E. *This Land, This South: An Environmental History.* Lexington: University Press of Kentucky, 1983.

"Crossroads: Congress, the Corps of Engineers and the Future of America's Water Resources." National Wildlife Federation and Taxpayers for Common Sense, March 2004.

Dahl, Thomas E. "Status and Trends of Wetlands in the Conterminous United States, 1998 to 2004." U.S. Department of the Interior, Fish and Wildlife Service, 2006.

———. "Wetlands Losses in the United States, 1780's to 1980's." U.S. Department of the Interior, Fish and Wildlife Service, 1990.

Dáte, S. V. *Quiet Passion: A Biography of Senator Bob Graham.* New York: Tarcher/Penguin, 2004.

DeGrove, John M. Interview by Cynthia Barnett, University of Florida Samuel Proctor Oral History Program, December 1, 2001, and December 8, 2001.

———. *Land Growth and Politics.* Chicago: American Planning Association's Planners Press, 1984.

deHaven-Smith, Lance. *Controlling Florida's Development.* Wakefield, N.H.: Hollowbrook Publishing, 1991.

———. *Environmental Concern in Florida and the Nation.* Gainesville: University of Florida Press, 1991.

Derr, Mark. *Some Kind of Paradise: A Chronicle of Man and the Land in Florida.* Gainesville: University Press of Florida, 1998.

"Desalination and Water Purification Technology Roadmap," U.S. Bureau of Reclamation and Sandia National Laboratories, January 2003.

de Villiers, Marq, *Water: The Fate of Our Most Precious Resource.* New York: Houghton Mifflin, 2000.

Dinar, Ariel, ed. *The Political Economy of Water Pricing Reforms.* New York: Oxford University Press, 2000.

Dodrill, David E. *Selling the Dream: The Gulf American Corporation and the Building of Cape Coral, Florida.* Tuscaloosa: University of Alabama Press, 1993.

Douglas, Marjory Stoneman. *Florida: The Long Frontier.* New York: Harper and Row, 1967.

Dovaz, Michel. *Water: From the Big Bang to the Bottle.* New York: Assouline Publishing, 2000.

"Estimated Use of Water in the United States, 2000." U.S. Geological Survey Circular 1268, updated February 2005.

Evans, Harry B. *Water Distribution in Ancient Rome: The Evidence of Frontinus.* Ann Arbor: University of Michigan Press, 1994.
Fernald, Edward A., and Elizabeth D. Purdum. *Water Resources Atlas of Florida.* Tallahassee: Florida State University Institute of Science and Public Affairs, 1998.
Flippen, J. Brooks. *Nixon and the Environment.* Albuquerque: University of New Mexico Press, 2000.
"Florida's Agricultural Water Policy: Ensuring Resource Availability." Florida Department of Agriculture and Consumer Services, July 2003.
Foglesong, Richard E. *Married to the Mouse: Walt Disney World and Orlando.* New Haven: Yale University Press, 2001.
Foster, John T., Jr., and Sarah Whitmer Foster. *Beechers, Stowes, and Yankee Strangers: The Transformation of Florida.* Gainesville: University Press of Florida, 1999, 89.
Gannon, Michael. *Florida: A Short History.* Gainesville: University Press of Florida, 1993.
———, ed. *The New History of Florida.* Gainesville: University Press of Florida, 1996.
Getches, David H. *Water Law in a Nutshell.* St. Paul, Minn.: West Publishing, 1997.
Gleick, Peter H. *The World's Water, 2000–2001: The Biennial Report on Freshwater Resources* Washington, D.C.: Island Press, 2000.
———. *The World's Water, 2002–2003: The Biennial Report on Freshwater Resources.* Washington, D.C.: Island Press, 2002.
———. *The World's Water, 2004–2005: The Biennial Report on Freshwater Resources.* Washington, D.C.: Island Press, 2004.
Glennon, Robert. *Water Follies: Groundwater Pumping and the Fate of America's Fresh Waters.* Washington, D.C.: Island Press, 2002.
Goldman-Carter, Jan, Andrew Schock, Nancy A. Payton, and Ann W. Hauck. "Road to Ruin: How the U.S. Government Is Permitting the Destruction of the Western Everglades." National Wildlife Federation, Florida Wildlife Federation, Council of Civic Associations, November 2002.
Graham, Bob. Interview by Samuel Proctor, University of Florida Samuel Proctor Oral History Program, May 3, 1991; April 6, 1994; February 21, 1995; and January 26, 2001.
Grunwald, Michael. *The Swamp: The Everglades, Florida and the Politics of Paradise.* New York: Simon and Schuster, 2006.
Guy, Andrew, and Patty Cantrell. *Liquid Gold Rush.* Michigan Land Use Institute, 2001.
Hamann, Richard. "Wetlands Loss in South Florida and the Implementation of Section 404 of the Clean Water Act." Report to the Office of Technology Assessment Oceans and Environment Programs, Office of Technology Assessment, U.S. Congress, September 1982.
Harper, Francis, ed. *The Travels of William Bartram: Naturalist's Edition.* New Haven: Yale University Press, 1958.
Harris, Jonathan M. *Environmental and Natural Resources Economics.* Boston: Houghton Mifflin, 2002.
Hodge, A. Trevor. *Roman Aqueducts and Water Supply.* London: Gerald Duckworth and Co., 1992.
Hopping, Wade L. Interview by Cynthia Barnett, University of Florida Samuel Proctor Oral History Program, July 22, 2003.
"Improving Florida's Water Supply Management Structure: Ensuring and Sustaining

Environmentally Sound Water Supplies and Resources to Meet Current and Future Needs." Florida Council of 100, September 2003.

Kallina, Edmund F., Jr. *Claude Kirk and the Politics of Confrontation.* Gainesville: University Press of Florida, 1993.

"Keys to Florida's Future: Winning in a Competitive World." Final Report of the State Comprehensive Plan Committee to the State of Florida, February 1987.

Kirk, Claude R., Jr. Interview by Julian Pleasants, University of Florida Samuel Proctor Oral History Program, October 29, 1998.

Lamis, Alexander P. *The Two Party South.* New York: Oxford University Press, 1984.

Lanier, Sidney. *Florida: Its Scenery, Climate and History.* Philadelphia: J. B. Lippincott and Co., 1876.

Louv, Richard. *Last Child in the Woods: Saving Our Children from Nature-Deficit Disorder.* Chapel Hill: Algonquin Books, 2006.

Marella, Richard L. "Water Withdrawals, Use, Discharge and Trends in Florida, 2000." U.S. Geological Survey Scientific Investigations Report 2004–5151, December 2004.

Marella, Richard L., and Marian P. Berndt. "Water Withdrawals and Trends from the Floridan Aquifer System in the Southeastern United States, 1950 to 2000." U.S. Geological Survey Circular 1278, 2005.

Martinez, Robert. Interview by Julian Pleasants, University of Florida Samuel Proctor Oral History Program, March 23, 1999.

McCally, David. *The Everglades: An Environmental History.* Gainesville: University Press of Florida, 1999.

McDonald, Bernadette, and Douglas Jehl, eds. *Whose Water Is It? The Unquenchable Thirst of a Water-Hungry World.* Washington, D.C.: National Geographic Society, 2003.

McIver, Stuart B. *Death in the Everglades: The Murder of Guy Bradley, America's First Martyr to Environmentalism.* Gainesville: University Press of Florida, 2003.

Morris, Allen, ed. *The Florida Handbook, 2001–2002.* Tallahassee: Peninsular Publishing, 2001.

Morris, Allen, and Joan Perry Morris, eds. *The Florida Handbook, 2005–2006.* Tallahassee: Peninsular Publishing, 2005.

Nash, Roderick Frazier. *Wilderness and the American Mind.* New Haven: Yale University Press, 2001.

Odum, Howard T., Elisabeth C. Odum, and Mark T. Brown. *Environment and Society in Florida.* Boca Raton, Fla.: Lewis Publishers, 1997.

O'Sullivan, Maurice, Jr., and Jack C. Lane, eds. *The Florida Reader: Visions of Paradise from 1530 to the Present.* Sarasota, Fla.: Pineapple Press, 1991.

Pettigrew, Richard A. Interview by Julian Pleasants, University of Florida Samuel Proctor Oral History Program, May 23, 2001.

Postel, Sandra. *Last Oasis: Facing Water Scarcity.* New York: W. W. Norton and Co., 1997.

———. "Liquid Assets: The Critical Need to Safeguard Freshwater Ecosystems." Worldwatch Paper 170, July 2005.

Proctor, Samuel. *Napoleon Bonaparte Broward: Florida's Fighting Democrat.* Gainesville: University of Florida Press, 1950.

Purdum, Elizabeth D. *Florida Waters.* Published jointly by Florida's water management districts, April 2002.

Rand, Honey. *Water Wars: A Story of People, Politics and Power.* Philadelphia: Xlibris, 2003.

Randazzo, Anthony, and Douglas S. Jones. *The Geology of Florida.* Gainesville: University Press of Florida, 1997.

Reed, Nathaniel P. Interview by Julian Pleasants, University of Florida Samuel Proctor Oral History Program, November 2, 2000, and December 18, 2000.

Reisner, Marc. *Cadillac Desert: The American West and Its Disappearing Water.* New York: Penguin Books, 1993.

Rogers, Peter. "Water Resources in the Twentieth and One Half Century: 1950–2050." Water Resources Update: Reflections on a Century of Water Science and Policy 116, March 2000.

Scholz, John T., and Bruce Stiftel, eds. *Adaptive Governance and Water Conflict: New Institutions for Collaborative Planning.* Washington, D.C.: Resources for the Future, 2005.

Segerfeldt, Fredrik. *Water for Sale: How Business and the Market Can Resolve the World's Water Crisis.* Washington, D.C.: Cato Institute, 2005.

Shroder, Tom, and John Barry. *Seeing the Light: Wilderness and Salvation; A Photographer's Tale.* New York: Random House, 1995.

Simon, Paul. *Tapped Out: The Coming World Crisis in Water and What We Can Do about It.* New York: Welcome Rain Publishers, 1998.

Simpson, Charles Torrey. *Out of Doors in Florida: The Adventures of a Naturalist, Together with Essays of the Wild Life and Geology of the State.* Miami: E. B. Douglas, 1923.

"Sinkholes, West-Central Florida: A Link between Surface Water and Ground Water." Excerpt from U.S. Geological Survey Circular 1182, 1999.

Steinberg, Ted. *Acts of God: The Unnatural History of Natural Disaster in America.* New York: Oxford University Press, 2000.

Tebeau, Charles W. *A History of Florida.* Miami: University of Miami Press, 1971.

"The Value of Water: Concepts, Estimates and Applications for Water Managers." American Water Works Association Research Foundation, 2005.

Wallsten, Scott, and Katrina Kosec. "Public or Private Drinking Water? The Effects of Ownership and Benchmark Competition on U.S. Water System Regulatory Compliance and Household Water Expenditures." AEI-Brookings Joint Center for Regulatory Studies, March 2005.

Ward, Diane Raines. *Water Wars: Drought, Flood, Folly and the Politics of Thirst.* New York: Riverhead Books Penguin Putnam, 2002.

"Water: The Power, Promise and Turmoil of North America's Fresh Water." *National Geographic,* special edition, 184, no. 5A (1993): 1–120.

The Water Barons: How a Few Powerful Companies Are Privatizing Your Water. International Consortium of Investigative Journalists. Washington, D.C.: Public Integrity Books, 2003.

"Water for People, Water for Life—UN World Water Development Report." UNESCO, March 2003.

"Wetlands Protection: Corps of Engineers Does Not Have an Effective Oversight Approach to Ensure that Compensatory Mitigation Is Occurring." U.S. Government Accountability Office, report GAO-05-898, September 2005.

Whitman, Christine Todd. *It's My Party Too: The Battle for the Heart of the GOP and the Future of America.* New York: Penguin Press, 2005.

Winsberg, Morton D. *Florida Weather.* Second edition. Gainesville: University Press of Florida, 2003.

Worster, Donald. *Nature's Economy: A History of Ecological Ideas.* Cambridge: Cambridge University Press, 1985.

———. *A River Running West: The Life of John Wesley Powell.* New York: Oxford University Press, 2001.

———. *Rivers of Empire: Water, Aridity and the Growth of the American West.* New York: Oxford University Press, 1985.

Ziewitz, Kathryn, and June Wiaz. *Green Empire: The St. Joe Company and the Remaking of Florida's Panhandle.* Gainesville: University Press of Florida, 2004.

Zwick, Charles J. Interview by Cynthia Barnett, University of Florida Samuel Proctor Oral History Program, August 17, 2000.

Index

Abdullah, crown prince of Saudi Arabia, 179
Acceler, 8, 94
Adams, Ansel, *Sierra Nevada: The John Muir Trail,* 182
AEI-Brookings Joint Center for Regulatory Studies, 165
Afghanistan, 126, 151
Africa, 41, 42, 108, 164, 182
agriculture, 3, 8, 21, 38, 45, 49, 51, 54, 55, 56, 61, 94, 96, 109, 156, 185. *See also* Florida Department of Agriculture and Consumer Services
Aguas del Tunari, 163
Alabama, 9, 34, 43, 115, 117, 120, 121, 126, 174
Alaska, 78
Alcoa, 74
Allentown, PA, 132
Aluminum Company of America, 74
Amarillo, TX, 158
American River, 13
American Water-Pridesa LLC, 178
American Water Services, 178
American Water Works Association, 151, 155, 164
Anderson-Columbia, 83, 84, 85
Andover, MA, 81
Andronaco, Meg, 143
Animas–La Plata dam, 46, 47, 151
Annan, Kofi, 126
Annapolis, MD, 121
Apalachicola Bay, 9, 115, 117, 120, 121
Apalachicola-Chattahoochee-Flint basin, 119, 120
Apalachicola-Chattahoochee-Flint (ACF) Compact, 120, 161
Apalachicola River, 14, 115, 120, 121
Apollo Beach, FL, 176, 177
Appalachia, 48
Appalachian Trail, 114
Aqua America, 166
Aquafina, 138
Aquarius Water Trading and Transportation, 125

Index

aquifers, 2, 6, 9, 34, 56, 66, 103, 117, 127, 152, 157, 161–62, 169, 171–74. *See also* specific names of aquifers, e.g., Floridan Aquifer
Aquifer Storage and Recovery (ASR), 160, 171–74
Arctic, 63
Arctic National Wildlife Refuge, 79
Argentina, 160
Arizona, 3, 100, 101, 126; water, 33, 45, 118, 149, 151
Arkansas, 54, 67
Arkansas River, 113
Arnold, Lee, 104, 105, 110
Arnold, Tony, 166
Arvida, 74, 75, 76, 88
Asia, 42, 164; water supply, 41, 42, 52, 124, 125, 127
Asian green mussels, 178
Askew, Reubin O'Donovan, 54, 57, 69, 95, 96
Aspinall, Wayne, 46
Atlanta, GA, 115, 116, 117, 120, 148, 166; water, 9, 114, 115, 117, 120, 130, 138, 139, 140, 164, 165, 166, 169
Atlantic City, NJ, 26
Atlantic Civil, 51, 185
Atlantic Ocean, 12, 14, 25, 45, 52, 87, 145, 146, 175
Atteberry, David, 1, 3, 10
Atteberry, Vivian, 1, 3, 10, 11
Audubon of Florida, 96
Australia, 33; water supply, 42
Azurix, 159–60, 161, 167

Baars, Theo D., Jr., 56
Babcock Ranch, 92
Bahamas, 125
Ball, Ed, 88
Baltimore, MD, 26, 122
Barefoot Bay, FL, 26
Barlow, Maude, 160
Bartram, William, 14, 135; *Travels,* 135
Bechtel, 163
Belle Glade, FL, 9, 145, 146
Belli, Lawrence, 93
Berg, Sanford V., 151, 152, 155
Berndt, Marian, 34, 35
Beverage Marketing Corporation, 130
Big Bend Power Plant, 176, 177
Big Cypress National Preserve, 57, 181, 182
Big Fish Lake, 109
Big Spring, 132
Bill 1, TX, 96
Bill 444, FL, 97
Bioterrorism Act, 151
Biscayne Aquifer, 21, 74
Biscayne Bay Coastal Wetlands, 185
Blain, Buddy, 107
Blake, Nelson Manfred, 18–19
Blanton, Donna, 73
Bloomberg, Michael, 103
Bloxham, William, 17, 18
Boca Raton, FL, 111, 141
Body for Life, 105
Bolivia, 163, 184
Bonaparte, Napoleon, 172
Bonita Springs, FL, 186
Bonneville Dam, 46
Boston, MA, 5, 48, 153, 169, 177, 187
bottled water industry, 9, 36, 128–44, 152, 156
Bova, Anthony, 141
Bowdre, Karon, 121
BP Capital, 158
Brazil, 184
Brinkley, Christie, 57
Brooklyn, NY, 38
Brooksville, FL, 62
Brous, Maria, 140
Broward, Napoleon Bonaparte, 19, 172
Broward County, FL, 37, 74, 75, 76; Planning Council, 74
Brower, David, 95
Brown, J. E. "Buster," 96
Brown v. Board of Education, 80
Bryant, Farris, 104
Buescher, Kent, 91
Buford Dam, 114
Bulkhead Act, FL, 24, 28
Bumpers, Dale, 54
Bunker Hill, MA, 48

Index

Bureau of Economic and Business Research, University of Florida, 2, 169
Bush, Barbara, 81
Bush, Columba (aka Columba Garnica Gallo), 81
Bush, George H. W., 78, 81, 86, 95
Bush, George W., 8, 67, 78, 85, 86, 92, 96–97, 99, 151, 173
Bush, Jeb, 7, 8, 36, 41, 78, 81–83, 84, 85, 86, 88, 89, 90–91, 92–93, 94, 96, 99–100, 101, 103, 104–5, 109, 111, 112, 113, 120, 144, 159–60, 184, 186, 188
Bush, John Ellis, 81
Bush v. Gore, 184
Butcher, Clyde, 180–83
Butcher, Jackie, 181
Butcher, Niki, 180, 181
Butcher, Ted, 181

Cabela's, 79
California, 6, 13, 28, 32, 41, 45, 78, 85, 96, 101, 112, 167, 169, 177, 179–82. *See also* Southern California
  Department of Water Resources, 96
  desalination plants, 149, 176, 179
  levees, 96
  Metropolitan Water District of Southern California, 159
  Senate Committee on Natural Resources and Water, 96
  swimming pools, 34
  water supply, 3, 10, 33, 46, 52, 118, 119, 125, 159, 171, 179, 191
  water use, 32
  wildlife habitats, 46
Caloosahatchee River, 17, 26, 27, 40, 58
Camp Blanding, FL, 22
Canada, 122, 123, 131. *See also* Ontario; Quebec
Canadian Environmental Law Association, 124
Canadian River Municipal Water Authority, 158
Canal 38 (C-38), 49, 71, 184
Carnival, 54
Carpenter, Robert, 184, 186
Carr, Marjorie Harris, 95
Carson, Rachel, 25; *Silent Spring,* 25
Carter, Jimmy, 46, 54
Carter, Luther J., 24
CCDA Waters, 130
Center for Global Solutions, 42
Center for Ocean-Atmospheric Prediction Studies, 61
Central America, 42
Central and South Florida Flood Control District, 49
Central and South Florida Project, 49
Chain of Lakes, 57
Charles I, king, 121
Charleston, SC, 65
Charlotte Harbor, FL, 92, 168; Estuary, 171
Charlotte, NC, 139
Chattahoochee River, 9, 114–15, 117, 119, 120, 186
Chattahoochee Spring, 115
Chelette, Angela, 137, 138
Chesapeake Bay, 50, 71, 93, 121, 161, 184
Chicago, IL, 99, 112, 127
Chiefland, FL, 111
Chile, 42
Chiles, Lawton, 57, 91, 92, 93
China, 42, 45, 126
Citizens for Michigan's Future, 125
citrus industry, 56, 101, 113
Civil War, 16, 17, 116, 118, 122, 175
Clean Air Act, 28, 48, 78
Clean Water Act, 28, 78, 86, 150
Clearwater, FL, 104
Cleveland, OH, pumping stations and sewage spills, 155
Clinton, Bill, 86, 92, 93, 173
Coastal Living, 89
Coca-Cola Corporation, 115, 130. *See also* Dasani
Codina, Armando, 81
Codina Group, 88
Colburn, David R., 23, 79, 81, 106
Coleridge, Samuel Taylor, 135; "Kubla Khan," 135
Collier County, FL, 91

Colliers Arnold, 104
Collins, LeRoy, 80
Coloma, CA, 13
Colombia, 126
Colorado, 3, 46, 47, 59, 113, 151, 156; River, 3, 4, 6, 45, 113, 118, 119, 156, 184
Colorado State University, 59
Columbia, MS, 65
Comprehensive Everglades Restoration Plan, 8, 51, 92, 93, 174, 184–85
Comprehensive Planning Act, 55
Confederacy, 16, 17
Congressional Office of Technology Assessment, 57
Connecticut, 27
Connecticut River, 187
Cooke Commission on Water Resources, 170
Cooperative Extension Service, University of Florida, 39
Coral, Cape, FL, 25, 26, 27, 30
Corpus Christi, TX, 138
Council of Canadians, 160
Council of 100, 104, 106, 110, 111, 112, 151
Covanta, 177
Covanta Tampa Construction, 178
Cracker Barrel, 169
Creek Indians, 114
Cross Bar Ranch Well Field, 108, 110
Cross-Florida Barge Canal, 95
Croton and Catskill Watershed, 51
Crowley, Brian, 73
Cryor, Jean, 122
Crystal Springs, FL, 142, 191
Crystal Springs Natural Spring Water, 138, 140
Curry, Judith, 62, 64
Cypress Gardens, 91
Cyprus, 125

Dade County, FL, 68, 69, 80
Dallas Water Utilities, 155
Danone, 130
Dasani, 138
Dáte, S. V., 69; *Palm Beach Post,* 69
Davis, Arthur Vining, 74
Davis, Cameron, 127
Daza, Victor Hugo, 163
DeBenedictis, Nick, 166
Deer Park Spring Water, 129, 136, 139, 140, 141
DEET, 137
deHaven-Smith, Lance, 23
Delaware, 51, 122
Delaware River Basin Commission Compact, 122
Deltona Corporation, 24
Democratic Party, 79, 80
de Normandy, Alphonse René le Mire. *See* Normandy, Alphonse René le Mire de
Denver, CO, 4, 113, 116
Denver Post, 46
Derr, Mark, 17
"Desalination and Water Purification Technology Roadmap," 170
Destination Florida, 101, 102, 103, 112
de Tocqueville, Alexis. *See* Tocqueville, Alexis de
Detroit, MI, 168; pumping stations, 155
Detroit River, 138
"Developments of Regional Impact," 69
Disney, 28–30, 112
Disston, Hamilton, 13, 17–18, 19, 88
Dockery, Paula, 97
Domenici, Pete V., 169
Don CeSar, 106
Douglas, Marjory Stoneman, 25, 95, 186; *The Everglades: River of Grass,* 25
Douglas, T. O'Neal, 102
Down, Barbara, 103
Draper, Eric, 96
drinking water, 3, 5, 31, 41, 45, 49, 52, 67, 74, 93, 117, 121, 138, 141, 142, 151, 154, 155, 158, 159, 173, 175, 189, 191. *See also* Safe Drinking Water Act
  standards, 32, 35, 139–40, 146, 165, 173, 174
drought, 5, 6, 7, 9, 10, 26, 32, 36–40, 53, 54, 62, 96, 97, 102, 103, 106, 108, 113, 117, 119, 120–21, 125, 143, 151, 178–79
DS Waters of America, 130, 131, 138, 140

Ducks Unlimited, 87
Duke University, Nicholas School of the Environment and Earth Science, 62
Durango Herald, 47

East (U.S.), 3
  hurricanes, 63
  water supply, 3–7, 9, 113, 117–18, 119, 122, 123, 129, 158, 159, 169, 171, 184, 192
East Coast Railroad, 18, 145
Ecclestone, Llywd, 104
Edward, Jacques, 188
Ellis, Roy, 142
El Niño, 61
Endangered Species Act, U.S., 28, 48, 78
Edwards Aquifer, 158
Enron, 160, 167
EPA. *See* U.S. Environmental Protection Agency
Environmental Systems Engineering Institute, 141
Ethiopia, 126
Europe, 26, 31, 42, 164
Everest, Mount, 138
Everest Water, 138
Everglades, 8, 9, 13, 15, 17, 21, 25, 51, 53, 67, 68, 70, 74, 78, 87, 93, 94, 161, 168, 181–83
  Aquifer Storage and Recovery (ASR), 174–75
  canals, 48, 49, 58, 76
  Comprehensive Everglades Restoration Plan, 8, 74, 92, 93, 95, 174, 184–85
  draining, 13, 16, 18, 19, 49, 51, 58, 69
  levees, 49, 93
  pump stations, 49, 76
  restoration, 51, 76, 79, 82–83, 85–86, 92–93, 95, 160, 184–86
  "Save Our Everglades" program, 71
  sugar cane farming, 68, 69
  urban development, 95, 98, 113, 186
  water supply, 95, 96, 113
Everglades Foundation, 56, 93
Everglades National Park (FL), 53, 93, 94, 185
Evian, 152
Edwards Aquifer, 158

Fairfax County Water Authority, 121, 122
Falcone, Elisabeth, 111
Falls Lake, 5
Farrell, Mark D., 162
Feather River, 45
Federal Emergency Relief Administration, 65
Federal Food, Drug, and Cosmetic Act, 131
FEMA City, 169, 171
Fernandina Beach, FL, 61
Financial Times, 164
fish, 5, 153
Fisher, Carl, 20
Flagler, Henry Morrison, 18, 21, 112, 145
Flint River, 9, 115, 119, 120, 161
Flippen, J. Brooks, 94; *Nixon and the Environment,* 94
flood control, 21. *See also* Central and South Florida Flood Control District
Florida. *See also* Bill 444; citrus industry; Council of 100; Cross-Florida Barge Canal; "The Impacts of Florida's Mature Residents"; Northwest Florida Water Management District; Sunshine Laws; Sunshine State; specific locality, e.g., Orlando, FL
  desalination plants, 179
  Division of Food Safety, 130
  drought, 102, 103, 106, 108, 117, 120–21, 143, 151, 174, 178–79
  hurricanes, 6, 8, 21, 43, 48, 49, 61, 64–65, 88, 106, 111, 169, 187, 191
  population growth, 2, 5, 6–7, 13, 23, 32, 80, 86, 93, 101, 105, 134, 169, 178, 179
  Senate, 111, 112
  Senate Natural Resources Committee, 111
  swimming pools, 34
  water costs, 151
  water shortage, 6, 38, 53, 97
  water supply, 2, 33, 43, 58, 59–62, 108, 179
  water use, 32, 33, 34, 35, 41, 52, 56

Florida (*continued*)
  wetlands, 7, 10, 24, 29, 35, 40, 53–76, 87–90, 107, 108, 154, 184, 186, 188, 191
Florida, Central, 29, 30, 61, 98, 102, 138, 143. *See also* Central and South Florida Flood Control District; Disney
  sinkholes, 2, 7
  water quality, 141, 173
  water supply, 35, 171
  wildlife, 115
Florida, North, 35, 80, 83, 95, 111, 121, 125, 184
  water supply, 7, 106, 110, 129, 135, 138
  water use, 142
Florida, South, 49, 51, 57–58, 60, 68, 71, 75, 80–81, 94, 95, 98, 99, 102, 122, 168, 172, 174, 184, 185. *See also* Central and South Florida Flood Control District; Central and South Florida Project; Destination Florida; Governor's Commission for a Sustainable South Florida; Intracoastal Waterway; South Florida Water Management District; and other specific localities and businesses, e.g., Disney; Everglades
  canals, 48
  flood control, 50
  hurricanes, 64, 65, 187, 188
  water shortage, 36
  water supply, 43, 48, 53, 58, 60, 74, 94, 106, 121, 174
  water use, 76, 160
  wetlands, 15, 51, 55, 162
  wildlife, 73, 115, 181
Florida Air and Water Pollution Control Act, 28
Florida Atlantic University, 111
Florida Bay, 15
Florida Chamber of Commerce, 152
Florida Department of Agriculture and Consumer Services, 130
Florida Department of Community Affairs, 75
Florida Department of Environmental Protection, 73
Florida Environmental Land and Water Management Act, 54
Florida Forever, 57, 90, 91
Florida International University, 111
Floridan Aquifer, 34, 35, 83, 139, 174
Florida State University, 10, 43; Center for Ocean-Atmospheric Prediction Studies, 61
Florida Supreme Court, 16
Florida Trend, 139, 140
Fort Collins, FL 59
Fort Lauderdale Sun-Sentinel, 74
Fort Myers, FL, 26, 28, 40, 64, 87, 98, 99, 104, 181
Fort Zachary Taylor, 175
Foster, Stephen, 105; "Camptown Races," 105; "Oh! Susanna", 105
Fountain of Youth, 12, 30, 137
Fountain of Youth Mineral Water, 137
Franklin, Shirley, 117, 165
Frontinus, Sextus Julius, 31, 34; *De Aquaductu,* 34

Gainesville, FL, 16, 35
Gardner, Royal, 89
Gary, IN, 134
General Agreement on Tariffs and Trade (GATT), 124
General Development Corporation, 172
George, Lake, 14
Georgia, 54, 65, 91, 115, 116, 126
  environmental and health issues, 35
  water shortage, 120
  water supply, 9, 34, 43, 105, 114, 115, 117, 121, 123, 128
  water use, 35, 120, 185–86
Georgia Department of Natural Resources, 120
Georgia Tech
  School of Earth and Atmospheric Science, 62, 64
Glacier Clear Water, 138
Gleick, Peter H., 41, 96, 156, 187
Glen Canyon Dam, 45, 46, 114

Index

Glendening, Parris, 121, 122
global warming, 59, 62, 63, 64
Golden Gate, FL, 26
Golden Springs LLC, 137
Gore, Albert, Jr., 92. *See also Bush v. Gore*
Gorrie, John, 13
Goss, Porter, 186
Governor's Commission for a Sustainable South Florida, 82
Graham, Bob, 25, 69, 70–71, 92, 96, 182
Graham, Ernest "Cap," 68
Graham, Philip, 68
Graham Companies, 68–69
Grand Coulee, 46
Grandmaster Flash, "New York, New York," 106
Grand Rapids, MI, 133
Grand Teton Mountains, 77, 78
Granholm, Jennifer, 123, 133, 134
Great Atlantic Coastline Railroad, 18
Great Depression, 22, 23, 26, 44, 68, 88
Great Lakes, 7, 42, 50–51, 121, 123, 124, 125, 133, 134, 143, 184. *See also* individual lakes
  pumping, 52, 123, 124, 126, 127, 133–34, 144, 171
  water supply, 9, 42, 52, 123, 127, 170, 174
  wetlands, 87, 93
Great Lakes Basin Compact, 122–23, 127, 144
Greece, 125
Green Bay, WI, 174
Green Berets, 108
Greenland, 63
Greeneville, TN, 138
Greenstein, Steve, 58
Green Swamp, FL, 171
groundwater pumping, 2, 5, 6, 7, 9, 35, 43, 52, 66, 96, 106, 107, 108, 110, 119, 127, 138, 142, 143, 153, 159, 161, 162, 174, 177, 178
Grunwald, Michael, 71
Gulf American Corporation, 26, 27
Gulf Hammock, FL, 138
Gulf of Mexico, 3, 14, 24, 25, 26, 51, 79, 87, 105, 106, 110, 115, 118, 134, 189
Gulf Oil, 158
Gurney, Edward J., 80

Haiti, 147, 188
Hall, Noah, 123
Hamann, Richard, 57, 58
Hardin, Garrett, 152, 153; "Tragedy of the Commons," 152
Harris, Katherine, 73
Harvard University, 68, 99
Harvey, Richard, 94
Helen, GA, 114, 116
Hemphill, Gary, 130, 141
Henderson Wetlands Protection Act, 73, 74
Herd, Ken, 177, 178
Highway A1A, 22, 145
Highway 1, 33
Highway 93, 44
highways, 1, 2, 22, 24, 34, 36, 56, 65, 70, 96, 133, 153, 191. *See also specific highways,* e.g., Kings Highway
Hildon, 141
Hillsborough County, FL, 103, 106, 109, 161, 177
Hilsenbeck, Richard, 91
Hilton Head Island, SC, 35
Hobe Sound, FL, 26
Hodge, Trevor, 32
Hoffman, Alfred, 8, 98, 99–101, 104
Holton, A. Linwood, 54
Home Depot, 115
Hoover, Herbert, 170
Hoover Dam, 44, 45, 46, 49
Hoover Dike, Herbert, 48, 49
Hopping, Wade L., 80, 95
Houston, TX, 139, 160
Huron, Lake, 123
Hurricane Andrew, 65
Hurricane Camille, 65
Hurricane Charley, 169
Hurricane Donna, 65
Hurricane Hugo, 65

Hurricane Katrina, 7, 40, 64, 66–67, 68, 144, 185, 187
hurricanes, 3, 6, 7, 8, 21, 40, 43, 49, 59, 61, 64–68, 88, 106, 111, 144, 169, 185, 187, 191
Hurricane Wilma, 187, 191
Hydraulic Age, 53

Ice Mountain, 133
Ichetucknee River, 83, 84, 85
Ichetucknee Springs State Park, 83
Illinois, 1, 11, 41, 52–53
"Impacts of Florida's Mature Residents, The" (study), 100
Imperial Valley, CA, 45, 156
India, 42, 126, 156, 162
Indian River, 61
Indian River Lagoon, FL, 25, 185
Indian Trace, FL, 74
Indians, 15. *See also* specific Indian tribes, e.g., Creek Indians
Indonesia, 147
Industrial Age, 44, 52
Institute for the Study of Planet Earth, 63
Internal Improvement Fund, 16, 17
International Bottled Water Association, 139
International Consortium of Investigative Journalists, 164
Interstate 4, 102
Interstate 10, 128
Interstate 75, 128, 168, 171, 189
Interstate 94, 125
Interstate 95, 102
Interstate 595, 76
Intracoastal Waterway, 14
Ipswich River, 5, 153–54; Watershed Association (IRWA), 153–54
irrigation, 3, 4, 32, 38, 40, 46, 47, 52, 149, 152, 157, 187
Iraq, 71, 126
Israel-Palestine, 42, 126
Istanbul, 31
Italy, 31, 129

Jackson, Andrew, 14
Jackson, Lake, 96
Jackson Hole, WY, 77
Jacksonville, FL, 18, 22, 61, 64, 75, 88, 102, 106, 122, 138
Jakarta, Indonesia, 147
Johnson, Lyndon, 69, 80

Kansas, 3, 157
Karachi, Pakistan, 147
karst, 2, 142
Kay, Stephen, 139
Kennebunkport, ME, 78
Kennedy, Jack, 69
Kennedy, John F., 80, 176, 179
Kerr, Robert, 120
Key, V. O., Jr., 79
Keys, FL, 78, 90, 175; Key West, 65, 175, 176
King, Jim, 122
King, Ziba, 168
Kings Highway, 168, 170, 171
Kirk, Claude Roy, Jr., 27–28, 54, 80, 186
Kise, Kent, 140
Kissimmee River, 17, 49, 60, 71, 73, 184, 187
Kissengen Springs, FL, 35
Klamath River, 119
Koppel, Sol, 38–39

La Jolla, CA, 112
Lakeland Ledger, 37
Lake Michigan Federation, 127
lakes. *See* Big Fish Lake; Falls Lake; George; Great Lakes; Huron; Jackson; Lanier; Michigan; Okeechobee; Ontario; Powell; Superior
Land Conservation Act, 54
Land and Water Conservation Fund, 27
Landsat 5, 60
Lanier, Lake, 117, 120, 121
Lanier, Sidney, 18, 114, 116; *Florida: Its Scenery, Climate and History,* 18
La Quinta Inn, 169
Las Vegas, NV, 45, 126, 147
Latin America, 13, 76, 81, 164
Law of the Biggest Pump, 96

Index

Law of the River, 119
Lee, FL, 128, 142
Lee County, FL, 27, 92
Lehman, Patrick, 173
Leopold, Aldo, 79
levees, 39, 51, 66, 68, 93, 96
Levy County, FL, 110
Lincoln, NE, 5
Linsky, Ron, 146
Little Muskegon River, 133
Longino, Charles, 100
Los Angeles, CA, 4, 45, 156
Louisiana, 185, 187
  erosion, 51, 63
  Hurricane Katrina, 66, 185, 187
  levees, 68
  wetlands, 7, 63, 67, 68, 71, 93, 184
Louv, Richard, 189; *Last Child in the Woods,* 189
Lubbock, TX, 158

MacDill Air Force Base, 108
Macedonia, 126
Mackin, Kerry, 153
Madison, FL, 62; County, 128, 131, 132, 135, 142
Madison Blue Spring, FL, 128, 129, 136, 137, 142, 143
Maine, 78, 143
Major League Baseball, 148
Manatee County, FL, 161
Manatee Springs, FL, 35
Marco Island, 24, 100, 189
Marella, Richard, 34, 35
Marino, Dan, 76
Marshall, Arthur R., Jr., 71
Marshall, Curtis H., 60
Martinez, Bob, 80–81
Maryland, 121, 122, 123. *See also* Annapolis; Baltimore; Glendening, Parris; Port of Baltimore
Maryland Department of the Environment, 121
Massachusetts, 5, 81, 153, 154
Massachusetts Department of Environmental Protection, 153, 154
Matthews, Barbara, 103
Mazyck, David, 139, 140–41
McCain, Kathryn "Katie", 155
McCally, David, 15
Mecan River, 131
Mecan Springs, WI, 132
Mecca Farms, 113
Mecosta County, MI, 133
Mehan, G. Tracy, III, 150
Merrian, Chip, 174
Mesa Vista Ranch, TX, 157, 158
Mesa Water, 158
Metropolitan Water District of Southern California, 159
Mexico, Gulf of, 3, 24, 26, 51, 105, 106, 110, 115, 118, 134, 189
Miami, Australia, 33
Miami, FL, 13, 18, 20, 22, 33, 59, 88, 111, 116, 172, 188, 190
  environmental issues, 83
  hurricane devastation, 68
  politics, 81
  temperatures, 61
  water, 106, 125, 190
  wetlands, 51
Miami Beach, FL, 20, 21, 25
Miami-Dade County, 185
  crime, 84
  development and water, 94, 185, 186
Miami Dolphins, 84, 101
Miami Herald, 69, 85, 99
Miami Lakes, FL, 69, 70
Miami River, 17, 21, 38, 121; Commission, 191
Michigan, 68, 87, 123, 125, 133, 134, 144. *See also* Citizens for Michigan's Future; Detroit; Grand Rapids; Granholm, Jennifer; Mecosta County; Sanctuary Springs; Sault Ste. Marie
Michigan, Lake, 123, 127, 133, 174. *See also* Lake Michigan Federation
Michigan Citizens for Water Conservation, 133
Michigan Department of Environmental Quality, 133
Mickey Mouse, 29

Middle East, 176
Midwest, 2, 7, 52, 62, 102, 131, 132–33, 155
Millennium Development Goals, 41
Miller, Sarah, 124
Milwaukee, WI, 155
Milwaukee Journal-Sentinel, 132
Mirasol, FL, 186
Mississippi, 5, 65
  Hurricane Katrina, 66, 185, 187
  water supply, 34
Mississippi River
  environmental issues, 51
  levees, 51
  water supply, 52, 123
Missouri River, 51, 119
Mobile, AL, 65
Moeur, Benjamin, 118
Molligan, Michael, 162
Montgomery County, MD, 122
Monsees, Steve, 108
Monticello, FL 62
Moran, Thomas, 182
Morgan, J. P., 158
Mormino, Gary, 22
Morse, Gary, 104
Morton, Rogers, 77–78
Moscow, Russia, 55
Murrow, Edward, R., 145; *Harvest of Shame,* 145

Naples, FL, 40, 64, 87, 98, 100, 168
National Academy of Sciences, 62
National Aeronautics Space Administration (NASA), 60
National Association of Evangelicals, 63
National Drought Mitigation Center, 5
National Hurricane Center, 3, 111
National Oceanic and Atmospheric Administration (NOAA), 59, 60
National Research Council, 174
National Testing Laboratories, 139
National Water Research Institute, 146
National Wildlife Federation, 87, 123
Nature Conservancy, 91
Nebraska, 3. *See also* University of Nebraska
Nelson, Bill, 186
Nepal, 126
Nessie Curve, 150, 151
Nestlé, 130, 131, 132, 133, 134, 137, 142, 143
  Deer Park Spring Water, 129, 136, 139, 140, 141
  Pure Life, 129, 136
  Nestlé SA, 130
  Nestlé Waters North America, 129, 132
Nevada, 3, 44, 47, 171
New England, 65
New Hampshire, 56, 93
New Haven, WI, 132
New Jersey, 5, 66, 85, 103, 122, 173, 177
Newlands, Francis G., 47
Newlands Act, 47
New Mexico, 3, 33, 157, 169
New Orleans, LA, 51, 66, 67, 106, 187
  Hurricane Katrina, 7, 40
  wetlands, 7, 40
New River, 20
New York, 6, 13. 38, 52, 85, 103, 116, 130, 155, 175
  erosion, 50
  water supply, 122
New York City, NY, 51, 112
  water quality, 131
  water supply, 5
"New York, New York" (song), 106
New York Stock Exchange, 88
New York Times, 4, 17, 118, 164
Niagara Falls, NY, 127
Nichols, Bill, 103
Niger River, 42
Nile River, 42
Nixon, Richard, 27–28, 69, 77–78, 80, 82, 94–95. *See also* Flippen, J. Brooks
No Net Loss, 86, 87, 89
Nordic Water Supply, 125
Normandy, Alphonse René le Mire de, 175
North Africa, 45
North America, 42, 62, 159. *See also* specific countries, e.g., United States
  water supply, 42, 132, 147, 157, 164
North American Free Trade Agreement (NAFTA), 124, 129

North Carolina, 5, 62, 66, 100, 102. *See also* Charlotte, NC; Outer Banks; Raleigh
erosion, 63
water supply, 5, 9, 66, 139
North Dakota, 3
Northeast (U.S.), 7, 62
Northern District of Alabama, 121
North Port, FL, 137, 172
North Shore, MA, 153
Northwest Florida Water Management District, 137
Norton, Gale, 86
Norway, 125
Nova Group, 124–25, 127

Oakwood Village, 38; Homeowners Association, 39
Obmascik, Mike, 46
O'Brien, James, 61
Ocala, FL, 62, 138
ocean levels, 62, 63
Ogallala Aquifer, 157
Ohio, 52, 139
Okeechobee, Lake, 15, 21, 40, 48, 49, 56, 71, 92, 160
pumping, 17, 58, 61, 94, 139
water supply, 53
Okefenokee Swamp, GA, 105
Oklahoma, 3, 157, 158
Olivera, Oscar, 163
Olmstead, Sheila Cavanagh, 148, 149
Ontario, 122, 123, 124, 125, 127
Ontario, Lake, 123
Orange County, FL, 29
orange industry. *See* citrus industry
Oregon, 119
Oregon State University, 118
Orlando, FL, 1, 13, 56, 61, 103, 104, 131, 182
drainage wells, 35
environmental issues, 64
water supply, 139
weather, 43
wetlands, 15, 49
Orlando Sentinel, 73
Oroville Dam, 45
Osceola County, FL, 29
Outer Banks, NC, 63
Overpeck, Jonathan T., 63
oyster industry, 50, 115, 117, 121

Pacific Institute for Studies in Development, Environment, and Security, 41, 96, 126, 156, 167
Pakistan, water issues, 126, 147, 162
Palm Beach, Australia, 33
Palm Beach County, FL, 9, 33, 76, 92, 98, 112–13, 145, 146, 147
Palm Beach Gardens, FL. *See* PGA National Resort of Palm Beach Gardens
Palm Beach Post, 69, 73
Panic of 1893, 19
Pantanal, 184
Panther Parkway, 74
Paraguay, 184
Parker Dam, 118
Pasco County, 106
sinkholes, 3
water supply, 24, 108, 109, 110, 177
Pataki, George, 85
Patel, Kiran, 41–42
Pawlenty, Tim, 85
PBS&J, 167
Peace River/Manasota Regional Water Supply Authority, 171, 172, 173
Pearl Harbor, 22
Pelican Island, FL, 25
Penney, J.C., 180, 182
Pennsuco, 68
Pennsylvania, 122, 132, 166
Pennsylvania Sugar Company, 68
Pensacola, FL, 54, 56, 83
Pepsi, 138
Perdido Key, FL, 90
Peres, Shimon, 126
Perge, Turkey, 32
Perrier Group, 129, 132, 133
Persian Gulf, 176
Persian Gulf War, 36
Peru, 42, 61

Index

PGA National Resort of Palm Beach Gardens, 104
Philadelphia, PA, 13, 19
Philippines, 126
Phillips Academy, 81
Phillips Petroleum, 158
Phoenix, AZ, 45, 154
Picayune Strand, FL, 91
Pickens, Boone, 9, 156, 157, 158, 159
Pielke, Roger, Sr., 59, 60, 62
Pinellas County, FL, 66, 106, 108, 109, 110, 177
Pittman, Craig, 87, 88
Plant, Henry B., 18
Plant City, FL, 103
Plonski, Ken, 101
Poland Spring, ME, 143
Polk County, FL, 37, 97
pollution, 24, 25, 28, 35, 51, 52, 57, 58, 62, 79, 85, 115, 120, 137, 170, 183
Pompeii, 66
Ponce de León, Juan, 12, 30
Popular Photography, 182
"Pork Chop Gang," 80
Port-au-Prince, Haiti, 147
Port of Baltimore, MD, 121
Portugal, 99
Poseidon Resources, 177, 178
Postel, Sandra, 161, 187
Potomac-Raritan-Magothy Aquifer, 5
Potomac River, 119, 121, 122
Powell, John Wesley, 3, 4, 6, 118
Powell, Lake, 4, 45, 46, 114
Pridesa S.A., 178
Public Citizen, 166
Public Trust Doctrine. *See* U.S. Public Trust
Public Utility Research Center, 151
Publix, 140
pumping stations, 76, 155
Purdum, Elizabeth D., 43; *Florida Waters,* 43

Quebec, 122, 127
Queensland, Australia, 33

rainfall, 2, 3, 4, 43, 58, 59–62, 108, 117, 118, 153, 169, 179
Raleigh, NC, 5
Rand, Honey, 107, 109
Randell, M. T. "Ted", 28
Randell Act, 28, 73
Ravan, Jack, 164
Reed, Nathaniel, 27, 28, 77–78, 186
Reedy Creek Improvement District, 29
Regional Atmospheric Modeling System (RAMS), 59, 60
Reisner, Marc, 53; *Cadillac Desert,* 6, 47
Remuda Ranch, FL, 26
Renaissance Vinoy, 106
Reno, NV, 46
Republican National Committee, 8, 99, 104
reverse osmosis, 138, 176
Revolutionary War, 48
Rio Grande River, 3, 118
Rio Rico, FL, 26
Riparian Rights Act, 16
Ristorante Bova, 141
Ritchie, Bruce, 36
River Ranch, FL, 26
Roanoke River, 119
Robinson, Michelle, 178
Rocky Mountain Institute, 187
Rodes, Charles Green, 20, 24
Roels, Harry, 164
Roman Empire, 31
Rome, 31, 33, 34, 44
Roosevelt, Franklin, 48, 182
Roosevelt, Theodore, 25, 47, 79, 82, 189, 191
Root, Lawrence C., 133
Rose Bowl, 25
Rosen, Jack, 25, 26, 40, 57
Rosen, Leonard, 25, 26, 40, 57
Royal Poinciana Hotel, 145
Ruhl, J. B., 119
Rule of Capture, 55, 96
Rummell, Peter, 90
RWE, 164

S. Pellegrino, 129

Sable, Cape, FL, 23
Sacramento, CA, 13
Safe Drinking Water Act, U.S., 131, 150, 165, 173
Sahara Desert, 27
saltwater, 5, 7, 34, 35, 53, 58, 63, 140, 142, 161, 169, 176
Salvation Army, 188
San Antonio, 158
Sanctuary Springs, MI, 133
San Francisco, CA, 45
San Francisco Bay, 3, 191
Sanibel, FL, 186
Sanibel Chamber of Commerce, 58
San Joaquin River, 3, 191
Santa Barbara, CA, 176
Sanvicente, Robin, 137
Sarasota, FL, 93
Sarasota County, FL, 161
Sarasota Herald-Tribune, 160
Saratoga of the West, 127
Saudi Arabia, 105, 176–77, 179
Sault Ste. Marie, MI, 124
Savannah, GA, 34, 35, 65
Savannah River, 119
"Save Our Everglades," 71, 182
Sawgrass Expressway, 74
Scarborough, Joe, 89
Schlesinger, William H., 62, 63
Schwarzenegger, Arnold, 85, 96
Science, 64
Scripps Research Institute, 112–13
Sears, 180
Segerfeldt, Fredrik, 162
Seibert, Steve, 109
Seminole Indians, 15
Senate. *See* Florida, Senate; Southern California; U.S. Senate
Serageldin, Ismail, 147
Shasta Dam, 46
Sherman, William Tecumseh, 115
Shula, Don, 101
Siemens, 164
Sierra Club, 95, 184
Silver Springs Bottled Water, 138
Simon, Paul, 41, 52; *Tapped Out,* 41
Simpson, Charles Torrey, 21; *Out of Doors in Florida,* 21
Sindelar, Roberta E., 132
sinkholes, 1–2, 3, 7, 10, 35, 36, 103, 107, 154
"Sinkholes" (brochure), 2
Sleep Inn, 169
Smid, Harry, 175
Smith, Adam, 144, 148; *The Wealth of Nations,* 148
Smith, Bob, 56, 93
Smith, Buckingham, 15, 16
Smith, Stanley K., 2, 169
Social Security, 150
South (U.S.), 4, 52, 69, 105
South America, 42. *See also* Colombia
South Carolina, 54, 100
  hurricanes, 65
  water supply, 9, 34, 174
  water use, 35
South Dakota, 3, 157
Southeast (U.S.), 139
  global warming, 63
  water quality, 139
  water supply, 120
Southern California, 159, 186
South Florida Water Management District, 51, 71, 87, 159, 186
Southwest Florida Water Management District, 107, 108, 162, 177
Spain, 12, 178
Spanish-American War, 175
Sports Illustrated, 57
Spragg, Terry, 125
springs, 2, 14, 21, 138, 192. *See also* Madison Blue Spring
St. Augustine, FL, 15
St. Joe Company (aka JOE), 88, 89, 110, 186
St. Johns River, 14, 18
St. Johns River Water Management District, 139
St. Lawrence River, 51, 52, 124
St. Lucie River, 17
St. Marks Wildlife Refuge, 73
St. Petersburg Times, 82, 87, 89, 105
State of the Union Address, 151

State Road 27, 83
Stegner, Wallace, 6
Stetson University, 89
Stevens, Danny, 110
Stevenson, Jim, 36
Steyaert, Louis T., 60
Stone and Webster Company, 177
Stowe, Harriett Beecher, 13, 18; *Palmetto Leaves,* 18; *Uncle Tom's Cabin,* 18
Struhs, David, 83, 84, 160
Sudan, 126
Suez and Veolia Environment, 164
Sugar industry, 16, 61, 68, 69, 82, 92, 145, 160
Sun Belt, 102
Sun City, AZ, 101
Sun City Center, FL, 99
Sunshine Laws, 162
Sunshine State, 2, 6, 8, 11, 22, 28, 30, 33, 35, 38, 80, 83, 98, 101, 102, 112, 131, 181. *See also* Florida
Superior, Lake, 123, 124
Susquehanna River Basin Compact, 122
Suwannee Limestone, 173
Suwannee River, 9, 14, 83, 105, 110, 113, 121, 122, 130, 181
Suwannee River Water Management District, 129, 130
Swamp and Overflowed Lands Act, U.S., 16, 17
swamps, 7, 8, 10, 13, 15, 16, 17, 18, 19, 21, 24, 25, 29, 53, 63, 67, 68, 69, 74, 75, 76, 79, 86, 87, 93, 100, 105, 115, 131, 171, 182, 183, 188
swimming pools, 5, 22, 34, 94, 146, 153
Syria, 42

Tallahassee, FL, 8, 24, 61, 67, 73, 83, 88, 104, 112, 119, 182
Tallahassee Democrat, 36
Tampa, FL, 22, 36, 41, 42, 62, 80, 99, 102, 103, 104, 108, 125, 142, 175
Tampa Bay, 38, 57, 106, 107, 108, 109, 110, 154, 175, 176, 177, 178, 189
Tampa Bay Seawater Desalination Plant, 178
Tampa Bay Water, 109, 177, 178
Tampa Bay Water Wars, 106, 110, 154
Tampa Electric Company, 176
Taylor, James, 141
Tellico Dam, 48
Tennessee River, 48
Tennessee Valley Authority (TVA), 48, 49
Texas, 6, 13, 179. *See also* Amarillo; Bill 1; Corpus Christi; Dallas Water Utilities; Houston; Lubbock; Mesa Vista Ranch
  desalination plants, 179
  Supreme Court, 96
  water supply, 179
  water use, 32, 52
Thames Water, 164
Thatcher, Margaret, 164
Thompson, Tommy G., 132
Three Sisters Springs, FL, 142
Time, 54, 64
Tolkien, J. R. R., 105
Tocqueville, Alexis de, 15–16; *Democracy in America,* 16
Toth, Lou, 71
Tragedy of the Commons, 152, 154, 157
Tucson, AZ
  drought, 151
  water costs, 151
Turanchik, Ed, 109
Turkey, 42, 125. *See also* Istanbul; Perge
Turner, Frederick Jackson, 45, 189; "The Significance of the Frontier in American History," 45
Twitchell, James, 141; "Living it Up: America's Love Affair with Luxury Goods," 141
Ty Nant, 141

U.S. Army, 108
U.S. Army Corps of Engineers, 47, 48, 49, 66, 68, 71, 75, 87, 89, 113, 114, 117, 123, 174, 184, 185, 186
U.S. Bureau of Reclamation, 47, 119, 170
U.S. Congress, 3, 16, 28, 46, 47, 48, 49, 52, 65, 66, 81, 86, 88, 92, 93, 94, 118, 119, 120, 123, 124, 127, 149, 151, 184, 185

Index

U.S. Court of Appeals
  Eleventh Circuit, 120
U.S. Department of Defense, 132
U.S. Department of Interior, 47
U.S. Environmental Protection Agency (EPA), 28, 35, 78, 85, 94, 117, 131, 140–41, 150, 171, 172, 173, 186, 187
  Office of Water, 150
U.S. Fish and Wildlife Service, 55, 71, 87
U.S. Food and Drug Administration, 131, 132, 138, 140
U.S. Geological Survey, 3, 34, 60, 137
U.S. House of Representatives, 52, 55, 185
  Interior Committee, 46
U.S. Public Trust Doctrine, 16
U.S. Secret Service, 77, 78
U.S. Senate, 69, 71, 80, 185, 192
  Energy and Natural Resources Committee, 169
  Natural Resources Committee, 93
Union (U.S.), 13, 175
United Kingdom, 160, 164
United Nations, 41, 126, 147
United States, 26, 32, 33, 45, 90, 91, 119, 129, 141, 145, 146, 150, 167, 182. *See also* specific localities, e.g., Great Lakes; specific states, e.g., Florida
  water costs, 147–48, 165–66
  water shortage, 5, 36, 38, 53, 97, 124, 127, 169
  water supply, 40, 42, 44, 47, 51, 52, 105, 118, 123–24, 126, 143, 149, 156, 157, 159, 163, 164, 169–70
  water use, 2, 32–33, 134, 187
  weather, 59, 64–66
United Water Resources, 164, 165
University of Arizona, Institute for the Study of Planet Earth, 63
University of Central Florida, Environmental Systems Engineering Institute, 141
University of Florida, 57, 139
  Bureau of Economic and Business Research, 169
  Cooperative Extension Service, 39
  Economic and Business Research, 2
  Public Utility Research Center Water Studies, 151
University of Louisville, Brandeis School of Law, 166
University of Nebraska, 5
University of South Florida, Center for Global Solutions, 42
University of Texas, Austin, 81
Unocal, 158
UPS, 115
US Filter, 164
Utah, 3, 125

Valdosta, GA, 91
Verizon, 54
Vergara, Sonny, 37
Villages, FL, 103, 104
Virginia, 54, 102, 121. *See also* Fairfax County Water Authority
  water supply, 121, 123
  water use, 122
  wetlands, 87
Vivendi, 164
Volusia Blue, FL, 137

Waite, Matthew, 87, 88
Wall Street, 98
Wal-Mart, 168, 188, 191
Walt Disney World. *See* Disney
Warm Mineral Springs, 137
Warren S. Henderson Wetlands Protection Act, 73, 74
Washington, 8, 52
Washington, D.C., 121, 182
Washington Post, 47, 68, 71, 94, 122
water, bottled. *See* bottled water
water, drinking. *See* drinking water
Water-Diamond Paradox, 148
water pumps, electric, 118, 157
Water Resource Associates of Tampa, 162
Water Resources Act, 55
Water Resources Development Act (WRDA), 124, 185
water shortage, 5, 6, 9, 32, 36, 38, 41, 42, 52, 53, 97, 120, 123, 124, 159
Water Wonderland, FL, 25, 26, 30

Waukesha, WI, 126–27
Waukesha County, WI, 127
"Way Down Upon the Suwannee River," 106
WCI Communities, 8, 98–101, 103
Weather Channel, 6, 43
Webb, Del, 101
Wekiva Springs, FL, 138
West (U.S.), 3, 5, 6, 7, 9, 52, 113, 118, 158, 193
West, John, 54
West Coast Regional Water Authority, 109
West Indies, 16
Weston, FL, 74, 75, 76, 88
West Palm Beach, FL, 139, 175
West Point Dam, 115
wetlands, 7, 10, 15, 16, 17, 24, 25, 29, 35, 40, 50, 52–76, 85–89, 107, 149, 154, 162, 184, 185, 186, 188, 191
White House, 92, 94, 95, 99
White Springs, FL, 142
Whitman, Christine Todd, 85, 173
Wilcox, J. Mark, 65
Wild Adventures, 91
wildfires, 36, 58, 102
wildlife, 5, 10, 15, 24, 25, 27, 46, 51, 58, 67, 73, 108, 142, 161, 170. *See also* National Wildlife Federation
wildlife refuge, 25, 67. *See also* Arctic Wildlife Refuge; St. Marks Wildlife Refuge, 170
Wilhite, Don, 5
Willard, Deb, 60
Willis, Brad, 143
Winsberg, Morton, 33; *Florida Weather,* 33
Winter Haven, FL, 91
Winter Park, FL, 1, 103
Wisconsin, 126, 133
  water quality, 130
  water supply, 123, 132, 155, 169
Wisconsin Department of Natural Resources, 132, 174
Withlacoochee River, 128
Woodraska, John "Woody," 159–60, 167
World Bank, 163
World Commission on Water for the 21st Century, 147
World Health Organization, 41
World War II, 22, 134
Worldwatch Institute, 161
Worster, Donald, 45
Worth, Lake, 145
Wyoming, 3, 77, 123

Yadkin-Pee Dee River, 119
Yale School of Forestry and Environmental Studies, 148

Zambezi River, 42
Zephyrhills Spring Water, 129, 136